职业教育双语教材
Bilingual Textbooks of Vocational Education

风光互补发电系统安装与调试

Installation and Commissioning of Wind-solar Complementary Power Generating System

姚 嵩 沈 洁 主编
Edited by Song Yao Jie Shen

李云梅 主审
Reviewed by Yunmei Li

本书是职业教育双语教材之一，由"鲁班工坊"主要建设单位天津轻工职业技术学院新能源类专业群教师团队，以"鲁班工坊"设备为载体，参照国家新能源专业相关教学标准编写而成，践行了党的二十大报告中提到的"深入推进能源革命"的相关要求，引导学生重视新能源的发展。

本书主要内容包括风光互补发电系统理论基础、功率跟踪技术、逆变器应用技术、风光互补发电实训系统应用等。书中印有二维码，配套有教学资源和视频。

本书可作为"鲁班工坊"的教学与培训用书，亦可作为职业院校光伏发电技术与应用、风力发电技术等新能源类专业教材。

图书在版编目（CIP）数据

风光互补发电系统安装与调试：汉英对照/姚嵩，沈洁主编.—北京：化学工业出版社，2019.12（2024.8重印）
职业教育双语教材
ISBN 978-7-122-35871-4

Ⅰ.①风⋯　Ⅱ.①姚⋯　②沈⋯　Ⅲ.①风力发电系统-双语教学-高等职业教育-教材-汉、英②太阳能发电-双语教学-高等职业教育-教材-汉、英　Ⅳ.①TM614②TM615

中国版本图书馆 CIP 数据核字（2019）第 279470 号

责任编辑：刘　哲　　　　　　　　　　装帧设计：韩　飞
责任校对：刘　颖

出版发行：化学工业出版社（北京市东城区青年湖南街13号　邮政编码100011）
印　　装：北京科印技术咨询服务有限公司数码印刷分部
787mm×1092mm　1/16　印张14¼　字数352千字　2024年8月北京第1版第5次印刷

购书咨询：010-64518888　　　　　　　售后服务：010-64518899
网　　址：http://www.cip.com.cn
凡购买本书，如有缺损质量问题，本社销售中心负责调换。

定　价：45.00元　　　　　　　　　　　　　　　　　版权所有　违者必究

前 言

为扩大与"一带一路"沿线国家的职业教育合作，贯彻落实天津市启动实施的将优秀职业教育成果输出国门与世界分享计划的要求，职业教育作为与制造业联系最紧密的一种教育形式，正在发挥着举足轻重的作用。根据共建"鲁班工坊"的合作意向，首批埃及来华培训教师在天津参加了"埃及·鲁班工坊EPIP师资研修班"，开展了为期一个月的师资培训。在埃及建立的"鲁班工坊"为埃及新能源企业和当地的中国企业培养急需的技术技能人才，实现优质资源共享。为了配合埃及"鲁班工坊"的理论和实训教学，开展交流与合作，提高中国职业教育的国际影响力，创新职业院校国际合作模式，输出我国职业教育优秀资源，课题组编写了《风光互补发电系统安装与调试》。

教材采用了项目导向、任务驱动的理念，构建光伏发电技术和风力发电技术基础知识与实际仿真系统实训相结合的课程内容。全书共包含三部分：第一部分为风光互补发电技术，主要介绍风光互补发电的专业基础知识；第二部分为风光互补发电系统与结构，着重介绍了风光互补实训平台的基本参数；第三部分实训项目为全书的重点内容。其中项目一到项目三，重点掌握风光互补发电系统理论基础、追日跟踪理论与实操、组件伏安特性测试技术、组件与蓄电池选择方法；项目四到项目五主要完成测试风力机特性曲线，及光伏阵列最大功率跟踪算法；项目六到项目七主要学习离网逆变器使用、并网逆变器应用，以及并网逆变器参数设置与电能质量分析方法；项目八到项目十为风光互补发电实训系统应用，运行与调试光伏发电系统和风光互补发电系统，并设计能源监控管理系统组态。

本教材与新能源专业教学资源库配合使用，资源库中有相应的学习资料。本教材配有二维码，可以即扫即学。教材中部分章节设有拓展阅读，介绍了"中国储能技术""中国光伏行业""中国风电产业"等的发展概况，引导学生坚定发展新能源的信心，为推动能源技术革命贡献自己的一份力量，践行党的二十大报告中关于"深入推进能源革命"的相关要求。

本教材由姚嵩、沈洁主编，李云梅主审，姚嵩、沈洁、李娜、马思宁、孙艳、皮琳琳、王欣、李良君参加了编写工作。姚嵩进行了框架设计，马思宁负责整体汇稿，沈洁对整体框架和全部内容进行了审核把关。第一、第二部分由姚嵩、沈洁编写；第三部分中，沈洁编写项目一，姚嵩负责编写项目二，马思宁编写项目三和项目九，李良君编写项目四，皮琳琳编写项目五，孙艳编写项目六和项目七，王欣编写项目八，李娜编写项目十。

限于编者水平，书中定有不少疏漏之处，恳请读者批评指正。

编 者

目 录

第一部分　风光互补发电技术 1

第二部分　风光互补发电系统与结构 6

第三部分　实训项目 9

 项目一　安装调试光伏追日跟踪系统 9
 项目二　测试光伏组件伏安特性 17
 项目三　设计系统组件与蓄电池容量 24
 项目四　风力机特性仿真 29
 项目五　验证光伏阵列最大功率跟踪算法 41
 项目六　学习离网型逆变器原理并完成系统测试 48
 项目七　测试并网型逆变器工作原理并分析电能质量 54
 项目八　运行与调试光伏发电系统 76
 项目九　运行与调试风光互补发电系统 80
 项目十　能源监控管理系统组态设计 84

参考文献 102

第一部分　风光互补发电技术

一、光伏发电技术

1. 光伏发电的基本原理

太阳能是一种辐射能，它必须借助能量转换器才能变换成电能。把太阳能转换成电能的能量转换器，叫做太阳能电池组件。

太阳能电池工作原理的基础，是半导体 p-n 结的光生伏打效应。光生伏打效应就是当物体受到光照时，其体内的电荷分布状态发生变化而产生电动势和电流的一种效应。

可将半导体太阳能电池的发电过程概括成如下 4 点：

① 先是收集太阳光和其他光，使之照射到太阳能电池表面上；

② 太阳能电池吸收具有一定能量的光子，激发出非平衡载流子（光生载流子）——电子-空穴对，这些电子和空穴应有足够的寿命，在它们被分离之前不会复合消失；

③ 这些电性符号相反的光生载流子，在太阳能电池 p-n 结内建电场的作用下，电子-空穴对被分离，电子集中在一边，空穴集中在另一边，在 p-n 结两边产生异性电荷的积累，从而产生光生电动势，即光生电压；

④ 在太阳能电池 p-n 结的两侧引出电极并接上负载，则在外电路中即有光生电流通过，从而获得功率输出，这样太阳能电池就把太阳能（或其他光能）直接转换成了电能，实训装置采用单晶硅材料太阳能电池。

这就是 p-n 结型硅太阳能电池发电的基本过程，如图 1.0.1 所示。若把几十个、数百个太阳能电池单体串联、并联起来，封装成为太阳能电池组件，在太阳光的照射下，便可获得具有一定功率输出的电能。

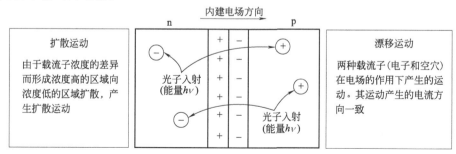

图 1.0.1　p 型、n 型材料结合扩散漂移产生内电场

2. 太阳能电池

(1) 太阳能电池的分类

按基本材料分类，太阳能电池有单晶体硅型、多晶体硅型、非晶硅型。

按结构分类，太阳能电池有同质结型、异质结型、肖特基结型、复合结型、液结型。
按用途分类，太阳能电池有空间型、地面型、光伏传感器。
按使用状态分类，太阳能电池有平板型、聚光型、分光型。
按封装材料分类，太阳能电池有刚性封装型、半刚性封装型、柔性衬底型。

（2）光伏发电的发展模式

独立运行光伏发电是相对于并网发电系统而言，可称作离网型光伏发电系统，其建设的主要目的是解决电网无法铺设的区域用电问题。偏远无电地区供电可靠性受气象环境、负荷等因素影响，供电稳定性相对较差，因此离网型光伏发电往往需要配备能量储存和能量管理设备。

并网型光伏发电系统可以将太阳能电池阵列输出的直流电转化为与电网电压同幅、同频、同相的交流电，实现与电网连接并向电网输电的功能。该发电系统较为灵活，但是由于日照的间歇性，因此对电网性能的要求较高，未来离网型光伏发电在电网中所占比例将会逐渐增加。

（3）光伏发电的发展趋势

① 离网型光伏发电系统将会在偏远地区、自然保护区广泛应用。
② 并网型光伏发电场所越来越灵活，屋顶使用率将会增加。
③ 光伏发电并网量增加，电网吸纳能力逐年增强。

二、风力发电技术

1. 风力发电的基本原理

风力发电是一个由风力机（风机）将捕获到的风能转化为机械能，并通过主轴、齿轮箱等传动机构将机械能传递给发电机，再由发电机将机械能转换为电能的过程，如图1.0.2所示。

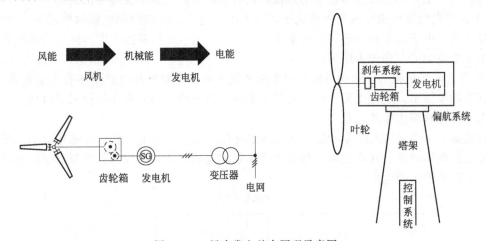

图1.0.2 风力发电基本原理示意图

由于兆瓦级风电机组齿轮箱损坏率较高，由此而有了直驱式风电机组（无齿轮箱）。风电机组常用功率调节方式有失速调节和变桨距调节两种。

2. 风力发电机组

（1）风力发电机组分类

按风轮叶片分类，风力发电机组有定桨型和变桨型。
按风轮转速分类，风力发电机组有定速型和变速型。
按传动机构分类，风力发电机组有齿轮箱升速型和直驱型。
按发电机分类，风力发电机组有异步型和同步型。

按并网方式分类，风力发电机组有并网型和离网型。

（2）风力发电的发展模式

陆地风力发电，其方向是低风速发电技术，主要机型是2～5MW的大型风力发电机组。这种模式关键是向电网输电。

近海风力发电，主要用于比较浅的近海海域，安装5MW以上的大型风力发电机，布置大规模的风力发电场。这种模式的主要制约因素是风力发电场的规划和建设成本。但是近海风力发电的优势是明显的，即不占用土地，海上风力资源较好。

（3）风力发电的发展趋势

① 变桨距调节方式迅速取代失速调节方式。
② 变速运行方式迅速取代恒速运行方式。
③ 机组规模向大型化发展。
④ 直驱永磁、异步双馈两种形式共同发展。

三、风光互补发电系统

风光互补发电系统由能量产生、存储和消耗三部分组成。太阳能、风能发电部分属于能量产生环节，将具有不确定性的太阳能、风能转化为稳定的能源。为了最大可能地消除由于天气等因素引起的能量供应与需求之间的不平衡，引入蓄电池来调节和平衡能量匹配，系统中的蓄电池用来承担能量的储存。能量消耗是指各种用电负载，有直流负载和交流负载两类。工作电压匹配的直流负载可以直接接入电路，工作电压不匹配的直流负载通过直流变换器后接入电路。交流负载接入电路时需要配备逆变器。风光互补发电系统如图1.0.3所示。通过太阳能光伏组件，将太阳能转换为直流电能，太阳能光伏组件安装方式采用双轴追日跟踪系统。光伏阵列有两个旋转自由度，可精确跟踪日光，保证光线垂直照射光伏组件。

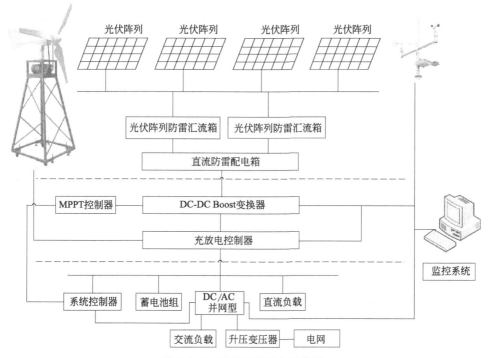

图1.0.3 风光互补发电系统图

为了减少光伏组件与控制器之间的连接线，方便维护，提高可靠性，一般需要在光伏组件与控制器之间增加直流汇流装置和直流配电箱。光伏阵列经汇流后可与 MPPT 控制系统相连，通过 MPPT 控制系统调整 DC-DC Boost 变换器的占空比（调节等效阻抗），实现最大功率跟踪。也可直接与直流负载相连，用于测试光伏特性。DC-DC Boost 变换器的输出端与智能充放电控制器相连，用于实现能量的存储。

通过变频器控制风速，输出功率经智能充放电控制器转换存储到蓄电池。蓄电池可直接带直流负载，或经过逆变装置处理转换为交流电能，再经变压器输入电网，实现并网发电。

监控系统可实时了解整个电站工作状态，实现控制与显示功能，如光照强度、环境温度、光伏电池数据监控，逆变器状态监控，发电数据监控等。

四、安全操作规范

为了顺利完成风光互补发电系统的实验项目，确保实验时设备安全、可靠及长期地运行，实验人员要严格遵守如下安全规程。

(1) 实验前的准备
① 实验前仔细阅读使用说明书，熟悉系统的相关部分。
② 实验前仔细阅读系统操作说明及实验的注意事项。
③ 实验前仔细阅读变频器用户手册，了解变频器的用法。
④ 实验前确保各系统控制柜电源处于断开状态。
⑤ 实验前根据实验指导书中相关内容熟悉此次实验的操作步骤。

(2) 实验中注意事项
① 严格按照正确的操作步骤给系统上电和断电，以免误操作给系统带来损坏。
② 在操作系统的过程中，能源转换存储控制系统蓄电池开关打开之后有一个等待智能充放电控制器自检初始化的过程，必须等到智能充放电控制器的"红灯"灭掉后才能进行下一步操作。
③ 在实验过程中，实验照明灯、模拟光源、电池板及周围的金属固件在光源的（长时间）照射下温度会上升，操作者不要用手指直接去触摸它们，以免烫伤。
④ 在实验过程中，有"危险"标志的地方为强电，注意安全。
⑤ 在实验过程中，模拟能源控制系统变频器的频率设置不能过高（不大于 20Hz）。

(3) 实验的进行步骤
① 预习报告详细完整，熟悉设备。实验开始前，指导老师要对学生的预习报告做检查，要求学生了解本次实验的目的、内容和安全实验操作步骤，只有满足此要求后，方能允许开始实验。

指导老师要对实验装置做详细介绍，学生必须熟悉该次实验所用的各种设备，明确这些设备的功能与使用方法。

② 建立小组，合理分工。每次实验都以小组为单位进行，每组由 2~3 人组成。
③ 试运行。在正式实验开始之前，先熟悉装置的操作，然后按一定安全操作规范接通电源，观察设备是否正常。如果设备出现异常，应立即切断电源，并排除故障；如果一切正常，即可正式开始实验。
④ 认真负责，实验有始有终。实验完毕后，应请指导老师检查实验资料。经指导老师认可后，按照安全操作步骤关闭所有电源，并把实验中所用的物品整理好，放回原位。

(4) 实验总结
这是实验的最后、最重要的阶段，应分析实验现象并撰写实验报告。每位实验参与者要独立完成一份实验报告，实验报告的编写应持严肃认真、实事求是的态度。

实验报告是根据实验中观察发现的问题，经过自己分析研究或组员之间分析讨论后写出的实验总结和心得体会，应简明扼要、字迹清楚、结论明确。

实验报告应包括以下内容：

① 实验名称、专业、班级、学号、姓名、同组者姓名、实验日期、室温等。

② 实验目的、实验内容、实验步骤。

③ 实验设备的型号、规格、铭牌数据及设备编号。

④ 实验资料的整理。

⑤ 用理论知识对实验结果进行分析总结，得出正确的结论。

⑥ 对实验中出现的现象、遇到的问题进行分析讨论，写出心得体会，并提出自己的建议和改进措施。

⑦ 实验报告应写在一定规格的报告纸上，保持整洁。

⑧ 每次实验每人独立完成一份报告，按时送交指导老师批阅。

第二部分 风光互补发电系统与结构

一、系统概况

本平台是针对高校新能源科学与工程等相关专业推出的开放式实验教学科研平台，由模拟光源跟踪装置、模拟风能装置、模拟能源控制系统、能源转换储存控制系统、并网逆变控制系统和能源监控管理系统六部分组成。平台采用模块化结构设计，可通过多种方式设计新能源应用系统。平台集电子信息、电力电子、自动控制等技术综合应用于一体，能满足高等院校电子信息工程、电气工程及其自动化、新能源科学与工程等相关专业的实验教学、工程设计和科研创新。该平台外形如下：

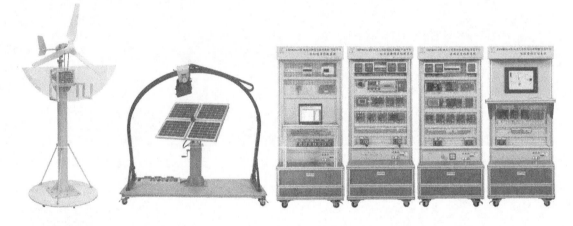

二、系统特点

① 模块化　采用工业标准，可根据不同的教学科研需求，组合不同的模块搭建不同的实验/开发系统。

② 开放性　软件和硬件部分资源全部向用户开放，用户也可根据实际需要增加新功能模块，开展探究型、创新型实验教学。

③ 新颖性　以微型电网新技术应用为向导，实验与设计开发相结合。

④ 先进性　系统涉及 PLC 控制、变频器调速、触摸屏、MCS51/PIC 单片机、DSP 处理器相关技术，具有跨专业、多学科融合的特点，可以满足不同层次用户需求。

三、技术性能

① 输入电源　三相四线，AC380V±10%，50Hz。

② 装置容量　<3kV·A。

③ 外形尺寸　1300mm×1100mm×2600mm（模拟风能装置）；2000mm×1500mm×2000mm（模拟光源跟踪装置）；800mm×600mm×1880mm（模拟能源控制系统）；800mm×600mm×1880mm（能源转换储存控制系统）；800mm×600mm×1880mm（并网逆变控制系统）；800mm×600mm×1880mm（能源监控管理系统）。

四、系统结构和组成

1. 风光互补发电技术实验/开发平台

该平台由计算机（用户自配）、模拟光源跟踪装置、模拟风能装置、模拟能源控制系统、能源转换储存控制系统、并网逆变控制系统和能源监控管理系统组成。

设备介绍

2. 模拟光源跟踪装置

该装置由4块太阳能电池组件、模拟光源（含灯具）、太阳能跟踪传感器、太阳能二维跟踪系统、模拟光源运行系统、蜗轮蜗杆减速箱、蜗轮蜗杆升降机、支架等组成。

① 模拟光源采用步进电机驱动，可在圆弧形的轨道上左右运行，模拟太阳运行轨迹，轨道的倾角可调节，模拟太阳光辐射角度。

② 4块太阳能电池组件固定安装在二维运动平台的支架上，中间装有太阳能跟踪传感器，底部采用蜗轮蜗杆升降机，可手动调节太阳能电池组件与模拟光源之间的距离。

③ 与模拟能源控制系统、能源转换储存控制系统组合，可完成光伏自动跟踪（基于传感器或经纬度）、太阳能控制器、最大功率跟踪（太阳能）等课题的研究。

3. 模拟风能装置

该装置由风力发电机、三相变频电机、编码器、传动装置、风机安全罩及塔架组成。

① 三相变频电机（带编码器）与风力发电机安装在塔架上，通过带轮传动，风叶旋转面装有半圆弧形透明有机玻璃材质的安全罩。

② 与模拟能源控制系统、能源转换储存控制系统组合，可完成风力机特性模拟、自然风模拟、最大功率跟踪（风能）、风光互补控制器、DC/DC变换器等课题的研究。

4. 模拟能源控制系统

该系统由控制屏（电源、网孔板、工具抽屉组成）、可编程序控制器（PLC）、编程线、模拟量模块、变频器、触摸屏、交流接触器、继电器、按钮、开关等组成。部件全部安装在网孔板上，硬件开放，组合灵活。与太阳能、风能装置组合，可完成光伏自动跟踪（基于传感器或经纬度）、风力机特性模拟、自然风模拟等课题的研究。

5. 能源转换储存控制系统

该系统由控制屏（电源、网孔板、工具抽屉组成）、光伏阵列汇流模块、直流电源防雷器、直流电压智能数显表、直流电流智能数显表、磁盘电阻器、断路器、开关电源、直流电压电流采集模块、CPU核心模块、人机交互模块、PWM驱动模块、通信模块、无线通信模块、温度告警模块、DC-DC Boost/Buck/Boost-Buck三种主电路模块、蓄电池组、充放电控制器、51 ISP下载器、PIC编程器等组成。

① 部件、模块全部安装在网孔板上，硬件开放，组合灵活。与太阳能、风能装置组合，可完成太阳能控制器、最大功率跟踪（太阳能、风能）、风光互补控制器、DC/DC变换器等课题的研究。

② 最大功率跟踪处理器采用 51 系列，支持在线下载，硬件开放，用户可以编写不同的 MPPT 算法以实现最大功率跟踪，并将调节参数发送给 PWM 驱动模块进行调节。PWM 驱动 CPU 采用 PIC 系列，接收调节参数，输出隔离的 PWM 驱动信号，控制主电路，实现功率调节。

6. 并网逆变控制系统

该系统由 DSP 核心模块、接口模块、液晶显示模块、键盘接口模块、驱动电路模块、Boost 电路模块、母线电压采样模块、电网电压采样模块、电流采样模块、温度告警模块、通信模块、开关电源、直流负载、交流负载、直流电压智能数显表、直流电流智能数显表、逆变输出电量表、隔离变压器、离网逆变器、DSP 仿真器等组成。

① 部件、模块全部安装在网孔板上，硬件开放，组合灵活，可开展 PWM 控制技术、离网逆变器、并网逆变器等课题的研究。

② 并网逆变将 DC24V 逆变成 AC36V、50Hz，经变压器升至 AC220V 与单相市电并网。主控制器采用 TI 定点 32 位 TMS320F2812 芯片，输出功率因数接近于 1。采用双闭环控制，内环为电流环，外环为电压环，并网同步采用数字锁相技术。

7. 能源监控管理系统

该系统由系统控制器核心模块、继电器模块、通信模块、15in❶ 工业平板电脑、键盘、鼠标、组态软件等组成。能源监控管理系统可与各控制系统通信，上位机软件可实时显示运行数据，并可根据控制要求自动或手动改变运行状态，开展能源监控系统方面的研究。

8. 风光互补发电技术视频教学软件

软件根据太阳能发电、风力发电、风光互补发电系统的知识点和技能点，采用视频教学方式，对系统的安装、接线、编程和调试过程进行讲练操作。包含以下视频教学内容：

① 模拟太阳能、模拟风能、能源转换储存控制、并网逆变控制、能源监控管理系统/器件介绍、安装、接线及安全教学视频；

② 太阳能自动跟踪、最大功率跟踪、并网逆变器教学视频；

③ PLC、变频器、触摸屏、MCS51/PIC 单片机、DSP 处理器、组态软件使用、编程教学视频。

9. 风光互补发电系统 3D 仿真实训软件

软件包括模拟风、模拟太阳光、风力发电场、风力发电机、多种类型的太阳能电池组件（单晶硅、多晶硅、非晶硅）、支架（固定式、单轴、双轴跟踪）、蓄电池组、风光互补控制器、逆变器（离网、并网）及适用于各种不同应用场合的交直流负载（交通灯、路灯、LED 屏、水泵等）仿真模型。软件以生动直观的仿真动画、3D 模型展示各个部件的结构和工作原理，可仿真多种发电系统（如独立光伏发电系统、独立风力发电系统、风光互补发电系统、离网发电系统、并网发电系统等）及应用系统实例（如太阳能交通灯、太阳能路灯、太阳能 LED 屏、太阳能水泵、风光互补路灯、风光互补监控等）。

风力发电场模拟多种真实场景，通过漫游、飞行模式，了解风电场布局以及大型风力发电机的运行状态。风力发电机机械组件 3D 模型逼真，支持任意方向旋转，可 360°全方位展示运行状态。

---

❶ 1in＝25.4mm。

第三部分 实训项目

项目一 安装调试光伏追日跟踪系统

【项目描述】

利用THWPFG-4型风光互补发电系统发电系统实训平台,学习模拟光源跟踪控制单元的基本知识,利用模拟能源控制单元,使太阳能电池板始终跟随模拟光源。

【能力目标】

① 学习三菱PLC的编程方法。
② 通过三菱PLC编程控制水平和俯仰运动机构,使太阳能电池板完成追日运动。
③ 通过分组学习,培养沟通与团队合作能力。

【项目环境】

完成该实训任务需要参考THWPFG-4型风光互补发电系统发电系统实训平台设备说明手册,学习模拟光源跟踪控制单元电气原理图,了解3ND583步进电机驱动器的电气指标及接口描述、M542步进驱动器使用原理、太阳能模拟追日跟踪传感器技术参数及接线原理、模拟光源技术参数、太阳能电池组件接线盒中接线方式。

模拟光源跟踪控制单元主要由太阳能电池组件、模拟光源、太阳能模拟追日跟踪传感器、太阳能电池板二维运动机构、步进电机、步进电机驱动器、减速箱、三菱可编程序控制器、按钮和继电器组成。模拟光源跟踪控制单元的电路框图如图3.1.1所示。

利用模拟能源控制单元的PLC控制灯光,模拟太阳东升西落的运行轨迹以及太阳光的入射角度;太阳能电池板上的模拟追日跟踪传感器,采集模拟光源的照度信息及位置信息,控制二维运动机构,使太阳能电池板始终跟随模拟光源。

模拟光源跟踪控制单元技术参数如下:
① 太阳能电池规格　20W/18V×4;
② 模拟光源功率　1000W;
③ 跟踪方式　双轴,俯仰180°,旋转360°;
④ 跟踪精度　<±1.5°;
⑤ 工作电压　DC24V;
⑥ 外形尺寸　2000mm×1200mm×2800mm。
模拟光源跟踪控制单元组成如下。

风光互补系统追(逐)日跟踪系统

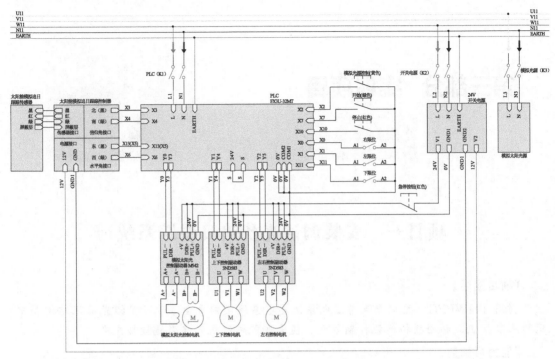

图 3.1.1 模拟光源跟踪控制单元电路框图

(1) 三菱可编程序控制器

三菱可编程序控制器输入、输出端定义如表 3.1.1 和表 3.1.2 所示。

表 3.1.1　PLC 输入端定义

类型	端子功能	三菱主机端口
输入	右限位开关	X0
	左限位开关	X1
	模拟光源控制按钮	X2
	向北传感器信号	X3
	向南传感器信号	X4
	向东传感器信号	X13
	向西传感器信号	X6
	开始按钮	X7
	停止按钮	X10
	下限位开关	X11

表 3.1.2　PLC 输出端定义

类型	端子功能	三菱主机端口
输出	模拟太阳光控制驱动器 PUL−	Y0
	上下控制驱动器 PUL−	Y1
	左右控制驱动器 PUL−	Y2
	模拟太阳光控制驱动器 DIR−	Y3
	上下控制驱动器 DIR−	Y4
	左右控制驱动器 DIR−	Y5

注：太阳能追踪传感器的俯仰角接口和水平角接口一端接 0V，另一端接主机输入端。

(2) 3ND583 步进电机驱动器

3ND583 步进电机驱动器是采用精密电流控制技术设计的高细分三相步进驱动器,适合驱动 57～86 机座号的三相步进电机。驱动器的电气指标如表 3.1.3 所示。

表 3.1.3　3ND583 驱动器电气指标

参　　数	最小值	典型值	最大值	单位
输出电流	2.1	—	8.3（均值 5.9）	A
输入电源电压	18	36	50	V DC
控制信号输入电流	7		16	mA
步进脉冲频率	0		400	kHz
脉冲低电平时间	1.2			μs
绝缘电阻	500			MΩ

3ND583 步进电机驱动器的接口描述,如表 3.1.4 和表 3.1.5 所示。

表 3.1.4　P1 端口（控制信号接口）描述

名称	功　　能
PUL+（+5V）	脉冲控制信号；脉冲上升沿有效；PUL－高电平时 4～5V,低电平时 0～0.5V。为了可靠响应脉冲信号,脉冲宽度应大于 1.2μs。采用时需串电阻
PUL－（PUL）	
DIR+（+5V）	方向信号：高/低电平信号。为保证电机可靠换向,方向信号应先于脉冲信号至少 5μs 建立。电机的初始运行方向与电机的接线有关,互换三相绕组 U、V、W 的任何两根线,可以改变电机初始运行的方向。DIR－高电平时 4～5V,低电平时 0～0.5V
DIR－（DIR）	
ENA+（+5V）	使能信号：此输入信号用于使能与禁止。ENA＋接+5V,ENA－接低电平（或内部光耦导通）时,驱动器将切断电机各相的电流,使电机处于自由状态,此时步进脉冲不被响应。当不需用此功能时,使能信号端悬空即可
ENA－（ENA）	

注：采用 24V 电源时,需要在控制器的信号端串入电阻 R,如图 3.1.7 PLC 控制原理图（见 14 页）所示。R 为 2kΩ、大于 1/8W 电阻。

表 3.1.5　P2 端口（功率接口）描述

名称	功　　能
GND	直流电源地
+V	直流电源正极,+18～+50V 间任何值均可,但推荐值+36V DC 左右
U	三相电机 U 相
V	三相电机 V 相
W	三相电机 W 相

驱动器的保护功能如下。

① 欠压保护　当直流电源电压＋V 低于 18V 时,驱动器绿灯灭,红灯闪烁,进入欠压保护状态。若输入电压继续下降至 16V 时,红、绿灯均会熄灭。当输入电压回升至 20V,驱动器会自动复位,进入正常工作状态。

② 过压保护　当直流电源电压＋V 超过 51V 时,保护电路动作,电源指示灯变红,保护功能启动。

③ 过电流和短路保护　电机接线线圈绕组短路或电机自身损坏时,保护电路动作,电源指示灯变红,保护功能启动。当过压、过流、短路保护功能启动时,电机轴失去自锁力,电源指示灯变红。若要恢复正常工作,需确认以上故障消除,然后电源重新上电,电源指示灯变绿,电机轴被锁紧,驱动器恢复正常。

(3) M542 步进驱动器

M542 型驱动器具有电机发热低、运行噪声低和运行平稳等特点，主要驱动 42、57 型两相混合式步进电机。建议工作电压范围为 24～36V DC，电压不超过 50V DC，不低于 20V DC。驱动器的使用说明如表 3.1.6 所示。

表 3.1.6 驱动器使用说明

驱动器功能	操 作 说 明
信号接口	PUL+和 PUL—为控制脉冲信号正端和负端；DIR+和 DIR—为方向信号的正端和负端；ENA+和 ENA—为使能信号的正端和负端
电机接口	A+和 A—接步进电机 A 相绕组的正端和负端；B+和 B—接步进电机 B 相绕组的正端和负端。当 A、B 两相绕组调换时，可使电机旋转方向反向
电源接口	采用直流电源供电，工作电压范围建议为 24～36V，电源功率大于 100W，电压不超过 50V DC 和不低于 20V DC
指示灯	驱动器有红绿两个指示灯，其中绿灯为电源指示灯，当驱动器上电后绿灯常亮；红灯为故障指示灯，当出现过压、过流故障时，故障灯常亮。故障清除后，红灯灭。当驱动器出现故障时，只有重新上电和重新使能才能清除故障
安装说明	驱动器既可卧式安装，也可立式安装。安装时应使其紧贴在金属机柜上以利于散热

(4) 太阳能模拟追日跟踪传感器

该传感器与控制器只检测传感器与模拟光源的位置偏差，并将位置信号转变成 4 个方向的开关量信号。当传感器对准模拟光源时，4 个信号均无输出。阴晴天调节电位器用来检测阳光强弱或白天晚上的信号。顺时针方向调节电位器听到继电器吸合时，再逆时针调节，听到继电器放开的声音。用手挡住传感器前端，继电器就会吸合。

太阳能模拟追日跟踪传感器技术参数如下：

① 系统供电　DC12V；

② 跟踪精度　1°；

③ 信号输出方式　无源触点；

④ 信号数量　4/5；

⑤ 输出信号定义　向左、向右、向上、向下/阴晴（白天/晚上）。

太阳能模拟追日跟踪传感器由阳光传感器（RY-CGQ-1-S）和控制器（RY-KZQ-D）两部分组成，如图 3.1.2 和图 3.1.3 所示。

图 3.1.2　阳光传感器（RY-CGQ-1-S）

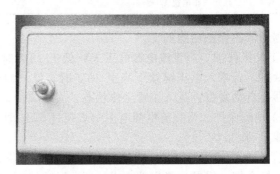

图 3.1.3　控制器（RY-KZQ-D）

控制盒出线定义如表 3.1.7 所示。

表 3.1.7 控制盒出线定义

名称	定义	备注
传感器接口	Line-1	传感器接口线按颜色一一对应,屏蔽层也要接好
电源接口	Line-2	
俯仰角接口	Line-3	
水平角接口	Line-4	

注：俯仰角接口、水平角接口一端接 0V，另一端接主机输入端。

电源接口	Line-2
棕(红)	DC12V+
蓝(黑)	DC12V-
黄	NC
绿	NC

俯仰角接口	Line-3
红(0V)	向北运动
黑(X3)	
黄(0V)	向南运动
绿(X4)	

水平角接口	Line-4
红(0V)	向东运动
黑(X13)	
黄(0V)	向西运动
绿(X6)	

(5) 模拟光源

模拟光源使用超亮型卤钨灯，用来模拟太阳光对太阳能电池组件照射产生电能。模拟光源如图 3.1.4 所示。

图 3.1.4 模拟光源

模拟光源技术参数如表 3.1.8 所示。

表 3.1.8 模拟光源技术参数

型号	QVF137
电压/频率	220V/50Hz
光源	PLUS PRO L 1000W R7s
灯座	R7s
质量/kg	1.75
迎风面积/mm²	0.054

模拟光源特征如下：
① 阳极氧化铝反射器，提供有效宽配光，更适合广泛光照明应用；
② 防水防尘等级达到IP54；
③ 高压铸铝灯体，抗腐蚀表面涂层；
④ 带铰链钢化玻璃，操作更加简便，无需维护。

(6) 太阳能电池组件

太阳能电池板是太阳能发电系统中的核心部分，也是太阳能发电系统中价值最高的部分。其作用是将太阳的辐射能转换为电能，或送往蓄电池中存储起来，或推动负载工作。太阳能电池组件如图3.1.5所示。

当有较多的太阳能电池组件串联组成电池方阵或电池方阵的一个支路时，需要在每块电池板的正负极输出端反向并联1个（或2~3个）二极管，这个并联在组件两端的二极管叫做旁路二极管。接线盒和旁路二极管如图3.1.6所示。

图3.1.5　太阳能电池组件

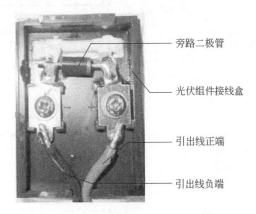

图3.1.6　接线盒和旁路二极管

【项目原理及基础知识】

(1) 光源模拟跟踪装置

该系统由光源模拟跟踪装置和光源模拟跟踪控制系统组成，包括太阳能电池组件、模拟太阳光灯、太阳能模拟追日跟踪传感器、太阳能板二维运动机构、步进电机、步进电机驱动

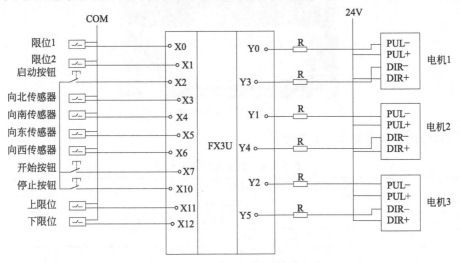

图3.1.7　PLC控制原理图

器、减速箱、三菱可编程序控制器、按钮和继电器等低压电器。

太阳能电池板组件的主要参数：标称功率 20W；工作电压 17.5V；工作电流 1.14A；开路电压 22.0V；短路电流 1.23A。

（2）PLC 控制原理图（图 3.1.7）

（3）控制柜接线图（图 3.1.8）

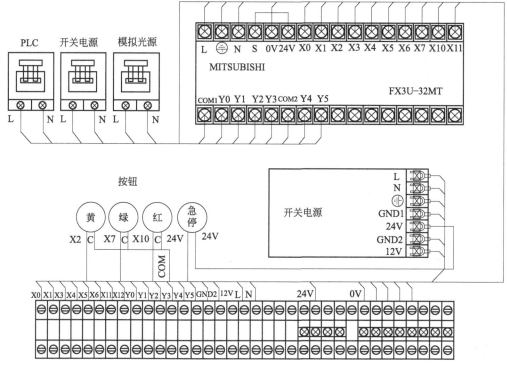

图 3.1.8　控制柜接线图

（4）步进电机驱动器接线图（图 3.1.9）

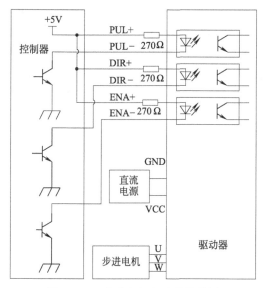

图 3.1.9　步进电机驱动器接线图

(5) 工作流程图（图 3.1.10）

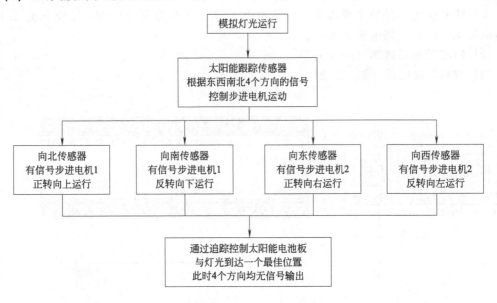

图 3.1.10　工作流程图

(6) 信号定义（表 3.1.1 和表 3.1.2）

太阳能电池板追日跟踪系统操作演示

【项目实施】

① 依次打开"模拟能源控制系统"的 PLC 开关、开关电源开关和模拟光源开关。

② 将 PLC 与电脑用 PLC 下载线连接，然后下载"太阳能电池板追日跟踪"示例程序。

③ 将 PLC 的"RUN/STOP"开关拨至"RUN"状态，系统将运行示例程序。

④ 按下追日系统控制开关的"开始"按钮（绿色按钮），各步进电机准备工作。

⑤ 按下"控制"按钮（黄色按钮），模拟光源开始运动；太阳能电池板上的传感器根据感光信号控制步进电机，使太阳能电池板跟随着模拟光源同步运动，模拟光源运动到左限位挡板后按原路返回到右侧限位挡板后一个周期完成。在不按下停止按钮前，模拟光源将重复周期性运动（"控制"按钮带自锁功能，按下后模拟光源才能运动）。

⑥ 按下"停止"按钮（红色按钮），所有步进电机停止运行。按下"开始"按钮后又继续运行。

⑦ 按下"急停"按钮，切断步进电机驱动器 24V 电源，松开后系统继续运行。

⑧ 实验结束后，按下"停止"按钮（红色按钮），所有步进电机停止运行。再依次关闭模拟能源控制系统的 PLC 开关、开关电源开关、模拟光源开关和总电源开关。如需进行后续实验，可不关闭总电源。

【项目作业】

① 画出太阳能电池板追日跟踪系统的硬件控制框图、连线图和 PLC 程序流程图。

② 根据 PLC 程序流程图编写太阳能电池板追日跟踪控制程序。

③ 模拟光源跟踪控制单元的核心是什么？

项目二　测试光伏组件伏安特性

【项目描述】

学习模拟光源跟踪控制单元的光源及电机等设备安装、接线原理及基本电气控制，学习能源转换储存控制单元的组成及电气原理图，熟悉电气控制步骤。

【能力目标】

① 理解太阳能电池板伏安特性。
② 掌握太阳能电池板最大功率的基本概念及测试方法。
③ 了解中国光伏行业发展概况，坚定科技自信。

中国光伏行业发展

光伏组件伏安特性测试

【项目环境】

完成该实训任务需要参考 THWPFG-4 型风光互补发电系统实训平台设备说明手册。

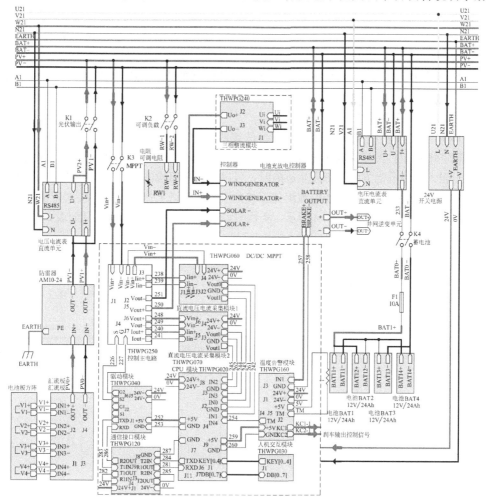

图 3.2.1　能源转换存储控制单元的电路框图

能源转换储存控制单元主要由光伏阵列汇流模块、直流电源防雷器模块、直流电压表、电流表、断路器、开关电源、MPPT 控制器、蓄电池组、智能充放电控制器组成，主要完成光伏汇流、防雷、电量测量、最大功率跟踪、储能和蓄电池管理等功能。能源转换存储控制单元的电路框图如图 3.2.1 所示。

各模块功能说明

① 光伏阵列汇流模块　用于把太阳能电池组件方阵的多路输出电缆集中输入、分组连接，不仅使连线井然有序，而且便于分组检查、维护。

② 直流电源防雷器模块　防雷器件是用于防止雷电浪涌电压侵入到太阳能电池方阵、交流逆变器、交流负载或电网的保护装置。

③ 直流电压、电流表　监测电量参数，如光伏输出电压、光伏输出电流等。

④ 断路器　主要完成各支路的断开与接入。

⑤ 连接模块　在被控对象与控制器之间起连接作用，方便插拔连接线。

⑥ 开关电源　为电路和设备提供稳定的直流电源。

⑦ MPPT 控制器　主要完成光伏、风机最大功率跟踪（MPPT）算法。

⑧ 蓄电池组　并网系统中附加的蓄电池，无风、无光照时可以供给电力，也可以在发电电力突变时起缓冲器作用、储存电能、电力调峰等扩大系统的使用范围，提高并网系统的附加经济值。

⑨ 智能充放电控制器　根据蓄电池电压高低，调节充电状态和电流的大小，防止蓄电池过充或过放，延长蓄电池使用寿命。

在本实训中着重了解光伏组件输出与可调负载之间的电气连线，并在实验中保证蓄电池与电路断开状态。下面详细介绍四个模块。

(1) 光伏阵列汇流模块

光伏阵列汇流模块将 4 组太阳能电池组件连线集中连接汇合后输出。

把太阳能电池组件的多路输出电缆集中、分组连接，不仅使连线井然有序，而且便于分组检查、维护。当太阳能电池方阵局部发生故障时，可以局部分离检修，不影响整体发电系统的连续工作。其具有过流保护和防太阳能电池组件反接的功能。4 路工作指示灯，使太阳能电池组件接入情况一目了然。其原理如图 3.2.2 所示。实物图如图 3.2.3 所示。端口定义如表 3.2.1 所示。

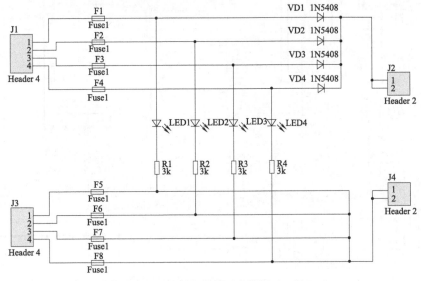

图 3.2.2　光伏阵列汇流模块原理图

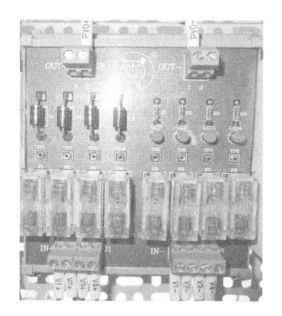
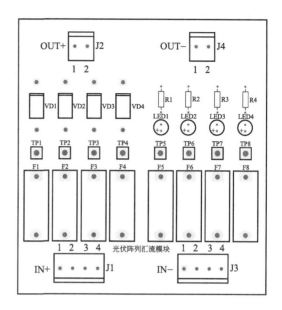

图 3.2.3 光伏阵列汇流模块

表 3.2.1 端口定义

序号	定义	说明	序号	定义	说明
1	J1:1(IN+)	第1路直流正输入	6	J3:3(IN−)	第3路直流负输入
2	J3:1(IN−)	第1路直流负输入	7	J1:4(IN+)	第4路直流正输入
3	J1:2(IN+)	第2路直流正输入	8	J3:4(IN−)	第4路直流负输入
4	J3:2(IN−)	第2路直流负输入	9	J2:1/2(OUT+)	直流汇流正输出
5	J1:3(IN+)	第3路直流正输入	10	J4:1/2(OUT−)	直流汇流负输出

(2) 直流电源防雷器模块

防雷器件是用于防止雷电浪涌电压侵入到太阳能电池方阵、交流逆变器、交流负载或电网的保护装置。为了保护太阳能电池方阵,每一个组件串中都要安装防雷器件。其采用温控断路技术,并内置过流保护电路,彻底避免防雷器自身发热引起的火险发生。工作状态及失效状态显示清晰直观。其原理如图 3.2.4 所示。技术指标如表 3.2.2 所示。实物如图 3.2.5 所示。端口定义如表 3.2.3 所示。

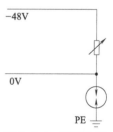

图 3.2.4 直流电源防雷器模块原理图

表 3.2.2 主要技术参数

产品型号	AM10~24
标称工作电压 U_n	24V DC
最大持续工作电压 U_c	36V DC
标称放电电流 I_n(8/20μs)	5kA
最大通流容量 I_{max}(8/20μs)	10kA

续表

保护水平 U_p（I_n时）	≤150V
响应时间	≤25ns
IP防护等级	IP20
阻燃等级，符合UL94	V0
接入导线面积	+、-、0线≥6mm²，地线≥10mm²
外形尺寸	单联：90mm×18mm×69mm；双联：90mm×36mm×69mm
工作环境	温度-40～+85℃，相对湿度≤95%（25℃），高度≤3km

图3.2.5 直流电源防雷器模块

表3.2.3 端口定义

序号	定义	说明	序号	定义	说明
1	IN+	直流电源正极输入	4	OUT+	直流电源正极输出
2	PE	地线（防雷地）(大地)	5	PE	地线（防雷地）(大地)
3	IN-	直流电源负极输入	6	OUT-	直流电源负极输出

(3) 直流单元模块

直流电压、电流表用来监测电量参数，如光伏输出电压、光伏输出电流等。

主要功能 直流电压（电流）测量、显示；RS-485、串行输出、模拟量输出和上下限告警、控制（接点输出）；可手动或上位机设定地址、波特率、显示数值、上下限报警阀值和回差。

技术指标特点 模块化设计，可根据用户需要选择单显示、显示加其他功能；采用最新PIC芯片，抗干扰能力强。

说明 安装结构为盘面安装，背后接线；外壳材料为阻燃塑料；隔离作用为电源、输入、输出相互隔离。

性能指标

① 最大显示：±19999

② 显示分辨率：0.001

③ 最大输入范围：电压DC 0～600V；电流DC 0～5A

④ 输入方式：单端输入

⑤ 精度等级：0.5级

⑥ 吸收功率：<1.2V·A

⑦ 测量速度：约5次/s

⑧ 输出负载能力：≤300Ω

⑨ 电源：AC220V±20%，50/60Hz

⑩ 工频耐压：电源/输入/输出间：AC 2kV/min，1mA

⑪ 脉冲串干扰（EFT）：2kV，5kHz
⑫ 浪涌冲击电压：2kV，1.2/50μs
⑬ 绝缘电阻：≥100MΩ

按键操作说明（表3.2.4）

表3.2.4 按键操作说明

显示字符	对应定义	数据范围
PASS	精度调节	0～9999
bAUD	波特率	1200、2400、4800、9600、19200
dISP	显示数值	0～19999
ADDr	本机地址	1～32
rHNu	继电器输出上限阀值	上限阀值＞下限阀值时，区间外告警
rLNu	继电器输出下限阀值	上限阀值＜下限阀值时，区间内告警 上限阀值＝下限阀值时，关闭告警功能
SAVE	保存退出	
E	不保存直接退出	
r-rE	回差	
DECi	小数点位置	

选定的功能不同，菜单也会有相应的增减。可操作按键有"SET""←""→"键。按"SET"键进入调试程序主菜单，LED显示PASS（精度调节）；按"←"键，显示dISP（显示数值）、DECi（单位和小数点位置）、SAVE（保存退出）、E（不保存直接退出）；按"→"键，反向循环显示。按"←"或"→"键到需要设定项，按"SET"键进入修改定项菜单。以下每项当调整到所需数值，按"SET"键退回到主菜单，详细说明如下。

① 显示数值（dISP）：LED显示00000（若已调整小数点显示位，则相应位置显示小数点），修改位闪动，按"→"键修改位加1，按"←"键修改位向左移动1位，如此循环，调至所需数值（显示数值对应满输入时显示数值，如输入DC100V，可设置显示为100.00）。

② 小数点位置（DECi）：LED显示当前小数点位置，按"→"键小数点右移1位，移至最后一位单位指示翻转（若无单位显示则忽略），按"←"键小数点左移1位，移至最高位单位指示翻转（若无单位显示则忽略）。

③ 本机地址（ADDr）：LED闪动显示本机地址，按"←"键数值加1，按"→"键数值减1，调至相应数值（1～32）。

④ 波特率（bAUD）：LED闪动显示当前设置波特率（1200～19200），按"←"键或"→"键更改数值。

⑤ 继电器动作阀值（rHNu/rLNu）：LED显示当前动作阀值，修改位闪动，按"→"键修改位加1，按"←"键修改位向左移动1位，如此循环，调至实际动作阀值。

⑥ 继电器动作回差设置（r-rE）：LED显示当前动作回差（回差值为实际测量的最后两位数值），修改位闪动，按"→"键修改位加1，按"←"键修改位向左移动1位，如此循环，调至实际动作回差。

⑦ 精度调节（PASS）：显示精度调整，修改精度时需相应的密码，出厂已调节好，无需用户调整。

当显示S时，按"SET"键保存所做的修改退出。显示E时，按"SET"键忽略修改直接退出。

说明：在任何一级菜单下，按键后，无操作时间大于60s系统将自动退出设置菜单。
实物图如图3.2.6所示。端口定义如表3.2.5和表3.2.6所示。

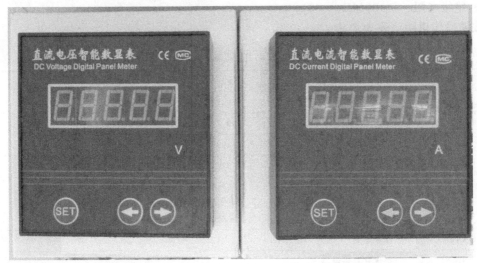

(a) 光伏输出电压　　　　　　　　　　　(b) 光伏输出电流

图 3.2.6　直流单元模块

表 3.2.5　直流电压表

序号	端口定义	说明
1	U+	被测电压输入
2	U−	
3	A	RS-485
4	B	
5	L	交流220V电源输入
6	N	

表 3.2.6　直流电流表

序号	端口定义	说明
1	I+	被测电流输入
2	I−	
3	A	RS-485
4	B	
5	L	交流220V电源输入
6	N	

（4）断路器开关顺序

注意：该实训是应用可调电阻（图3.2.7）作为负载，参照表3.2.7完成实训任务。

图 3.2.7　可调电阻

表 3.2.7 断路器开关顺序

序号	功能	断路器通电顺序	断路器断电顺序	可调电阻通电顺序	可调电阻断电顺序
1	光伏输出	1	3	1	2
2	可调负载	×	×	2	1
3	光伏 MPPT	2	2	×	×
4	蓄电池	3	1	×	×

本实训中使用的是旋转式可调变阻器，可以将光源和光伏电池板调到一个固定位置，按照阻值从大到小的顺序测量多组光伏输出电流、光伏输出电压。测得一组数据后，可以改变光源角度再进行一组测试，光源在不同角度时，比较伏安特性数据曲线的不同特性。

【项目原理及基础知识】

太阳能光伏发电的能量转换器是太阳能电池，又称为光伏电池。光伏电池在工作时，随着光照强度、环境温度和负载的不同，其端电压将发生变化，使输出功率也发生很大变化。光伏电池输出具有非线性特征，是一种不稳定的电源。本实训就是通过光伏电池在固定的光照强度下，测出当前环境的光伏电池的伏安特性，找出当前环境下光伏电池的最大输出功率点。

【项目实施】

① 合上"模拟能源控制系统"的"总电源"开关，系统得电，三相电源指示灯亮。

② 合上"模拟光源"空气开关，使模拟太阳光灯打开。

③ 合上"能源转换储存控制系统"的"总电源"开关，系统得电，三相电源指示灯亮。

光伏组件伏安特性测试操作演示

④ 记录光伏输出电压，此值即为光伏开路电压。

⑤ 合上"能源转换储存控制系统"的"光伏输出"和"可调负载"空气开关。

⑥ 调节可调电阻，按照阻值从大到小顺序，测量多组光伏输出电流、光伏输出电压，记录于表 3.2.8。

表 3.2.8 光伏电池的伏安特性测试表

序号	电压/V	电流/A	功率/W	备注
1	0			短路电流
2	1			
3	2			
4	3			
5	4			
6	5			
7	6			
8	7			
9	8			
10	9			
11	10			

续表

序号	电压/V	电流/A	功率/W	备注
12	11			
13	12			
14	13			
15	14			
16	15			
17	16			
18	17			
19	18			
20	19			开路电压

⑦ 通过记录的电压、电流数据，计算每个电压、电流对应的功率。

⑧ 实验结束后，依次关闭模拟能源控制系统的"模拟光源"空气开关和总电源开关，能源转换储存控制系统的"光伏输出""可调负载"空气开关和总电源开关。如需进行后续实验，可不关闭总电源。

【项目作业】

① 根据测量的数据，绘制 U-I 曲线、功率曲线，并找出最大功率点。
② 计算哪个点是最大功率输出点，为什么？

项目三　设计系统组件与蓄电池容量

【项目描述】

学习太阳能光伏发电技术的关键内容，了解太阳能电池组件和蓄电池选择的原理和基本知识。能够根据实际环境条件，计算太阳能电池组件的容量和数量，不同条件下合理设计蓄电池的容量及数量。

【能力目标】

① 根据要求计算太阳能电池组件的容量和数量。
② 根据要求计算蓄电池的容量和数量。
③ 了解储能技术发展概况，坚定投身光伏行业的信心。

中国储能技术
发展概况

【项目环境】

太阳能电池组件的设计原则是要满足平均天气条件（太阳辐射量）下负载每日用电量的需求，也就是说太阳能电池组件的全年发电量要等于负载全年用电量。因为天气条件有低于和高于平均值的情况，因此，设计太阳能电池组件是满足光照最差、太阳能辐射量最小季节的需要。如果只按平均值去设计，势必造成全年三分之一多时间的光照最差季节蓄电池的连续亏电。蓄电池长时间处于亏电状态，将造成蓄电池的极板硫酸盐化，使蓄电池的使用寿命和性能受到很大影响，整个系统的后续运行费用也将大幅度增加。设计时也不能考虑为了给蓄电池尽可能快地充满电而将太阳能电池组件设计得过大，这样在一年中的绝大部分时间里太阳能电池的发电量会远远大于负载的用电量,造成太阳能电池组件的浪费和系统整体成本的过高。因此,太阳能电池组件设计的最好办法,就是使太阳能电池组件能基本满足光照最差季节的需要,就是在光照最差的季节蓄电池也能够基本上天天充满电。

有些地区,最差季节的光照度远远低于全年平均值,如果还按最差情况设计太阳能电池组件的功率,那么在一年中的其他时候发电量就会远远超过实际所需,造成浪费。这时只能考虑适当加大蓄电池的设计容量,增加电能储存,使蓄电池处于浅放电状态,弥补光照最差季节发电量不足对蓄电池造成的伤害。有条件的地方还可以考虑采取风力发电与太阳能发电互相补充(简称风光互补)及市电互补等措施,达到系统整体综合成本效益的最佳。

(1) 太阳能电池组件及方阵的设计方法

设计和计算太阳能电池组件大小的基本方法,就是用负载平均每天所需要的用电量(单位:安时或瓦时)为基本数据,以当地太阳能辐射资源参数如峰值日照时数、年辐射总量等数据为参照,并结合一些相关因素数据或系数综合计算而得出的。

在设计和计算太阳能电池组件或组件方阵时,一般有两种方法。一种方法是根据上述各种数据直接计算出太阳能电池组件或方阵的功率,根据计算结果选配或定制相应功率的电池组件,进而得到电池组件的外形尺寸和安装尺寸等。这种方法一般适用于中、小型光伏发电系统的设计。另一种方法是先选定尺寸符合要求的电池组件,根据该组件峰值功率、峰值工作电流和日发电量等数据,结合上述数据进行设计计算,在计算中确定电池组件的串、并联数及总功率。这种方法适用于中、大型光伏发电系统的设计。下面以第二种方法为例介绍常用的太阳能电池组件的设计计算公式和方法。

(2) 太阳能电池组件及方阵的基本计算方法

计算太阳能电池组件的基本方法,是用负载平均每天所消耗的电量(Ah)除以选定的电池组件在一天中的平均发电量(Ah),算出整个系统需要并联的太阳能电池组件数量。这些组件的并联输出电流就是系统负载所需要的电流。具体公式为:

$$电池组件的并联数 = \frac{负载日平均用电量(Ah)}{组件日平均发电量(Ah)} \quad (3.3.1)$$

其中   组件日平均发电量=组件峰值工作电流(A)×峰值日照时数(h)

再将系统的工作电压除以太阳能电池组件的峰值工作电压,就可以算出太阳能电池组件的串联数量。这些电池组件串联后,就可以产生系统负载所需要的工作电压或蓄电池组的充电电压。具体公式为:

$$电池组件的串联数 = \frac{系统工作电压(V) \times 1.43}{组件峰值工作电压(V)} \quad (3.3.2)$$

系数 1.43 是太阳能电池组件峰值工作电压与系统工作电压的比值。例如,为工作电压 12V 的系统供电或充电,太阳能电池组件的峰值电压是 17~17.5V (12×1.43);为工作电压 24V 的系统供电或充电,峰值电压为 34~34.5V (24×1.43)等。因此为方便计算,用系统工作电压乘以 1.43 就是该组件或整个方阵的峰值电压近似值。例如,假设某光伏发电系统工作电压为 48V,选择了峰值工作电压为 17.0V 的电池组件,计算电池组件的串联数=48V×1.43/17.0V=4.03≈4(块)。

有了电池组件的并联数和串联数后,就可以很方便地计算出这个电池组件或方阵的总功率,计算公式是:

电池组件(方阵)总功率(W)=组件并联数×组件串联数×选定组件的峰值输出功率(W)

$$(3.3.3)$$

(3) 蓄电池和蓄电池组的设计方法

蓄电池的任务是在太阳能辐射量不足时，保证系统负载的正常用电。要能在几天内保证系统的正常工作，就需要在设计时引入一个气象条件参数：连续阴雨天数。一般计算时都是以当地最大连续阴雨天数为设计参数，但也要综合考虑负载对电源的要求。对于一般的负载，如太阳能路灯等，可根据经验或需要在3~7天内选取。对于重要的负载，如通信、导航、医院救治等，则在7~15天内选取。另外还要考虑光伏发电系统的安装地点，如果在偏远的地方，蓄电池容量要设计得较大，因为维护人员到达现场需要很长时间。实际应用中，有的移动通信基站由于山高路远，去一次很不方便，除了配置正常蓄电池组外，还要配备一组备用蓄电池组，以备不时之需。这种发电系统把可靠性放在第一位，已经不能单纯考虑经济性了。

蓄电池的设计主要包括蓄电池容量的设计计算和蓄电池组串并联组合的设计。在光伏发电系统中，大部分使用的都是铅酸蓄电池，主要是考虑到技术成熟和成本等因素，因此下面介绍的设计和计算方法也主要以铅酸蓄电池为主。

(4) 蓄电池和蓄电池组的基本计算方法

先将负载每天需要的用电量乘以根据当地气象资料或实际情况确定的连续阴雨天数，就可以得到初步的蓄电池容量，然后将得到的蓄电池容量数除以蓄电池容许的最大放电深度系数。由于铅酸蓄电池的特性，在确定的连续阴雨天内绝对不能100%地放电而把电用光，否则蓄电池会在很短的时间内寿终正寝，大大缩短使用寿命。因此需要除以最大放电深度系数，得到所需要的蓄电池容量。最大放电深度的选择，需要参考蓄电池生产厂家提供的性能参数资料。一般情况下，浅循环型蓄电池选用50%的放电深度，深循环型蓄电池选用75%的放电深度。计算蓄电池容量的基本公式为：

$$蓄电池容量 = \frac{负载日平均用电量(Ah) \times 连续阴雨天数}{最大放电深度} \qquad (3.3.4)$$

【项目原理及基础知识】

(1) 太阳能电池组件及方阵的计算相关因素的考虑

太阳能电池组件及方阵的基本计算方法完全是理想状态下的计算。根据上述计算公式计算出的电池组件容量，在实际应用当中是不能满足光伏发电系统的用电需求的。为了得到更准确的数据，就要把一些相关因素和数据考虑进来并纳入到计算中。

与太阳能电池组件发电量相关的主要因素有两点。

① 太阳能电池组件的功率衰降　在光伏发电系统的实际应用中，太阳能电池组件的输出功率（发电量）会因为各种内外因素的影响而衰减或降低。例如，灰尘的覆盖、组件自身功率的衰降、线路的损耗等各种不可量化的因素，在交流系统中还要考虑交流逆变器的转换效率因素。因此，设计时要将造成电池组件功率衰降的各种因素按10%的损耗计算，如果是交流光伏发电系统时，还要考虑交流逆变器转换效率的损失也按10%计算。这些实际上都是光伏发电系统设计时需要考虑的安全系数。设计时为电池组件留有合理余量，是系统年复一年长期正常运行的保证。

② 蓄电池的充放电损耗　在蓄电池的充放电过程中，太阳能电池产生的电流在转化储存的过程中会因为发热、电解水蒸发等产生一定的损耗，也就是说蓄电池的充电效率根据蓄电池的不同一般只有90%~95%。因此在设计时也要根据蓄电池的不同，将电池组件的功率增加5%~10%，以抵消蓄电池充放电过程中的耗散损失。

(2) 实用的太阳能电池组件及方阵计算公式

在考虑到各种因素的影响后，将相关系数纳入到式（3.3.1）和式（3.3.2）中，才是一个设计和计算太阳能电池组件的完整公式。

将负载日平均用电量除以蓄电池的充电效率,就增加了每天的负载用电量,实际上给出了电池组件需要负担的真正负载;将电池组件的损耗系数乘以组件的日平均发电量,这样就考虑了环境因素和组件自身衰降造成的组件发电量的减少,有了一个符合实际应用情况下的太阳能电池发电量的保守估算值。综合考虑以上因素,得出计算公式如下:

$$电池组件的并联数 = \frac{负载日平均用电量(Ah)}{组件日平均发电量(Ah) \times 充电效率系数 \times 组件损耗系数 \times 逆变器效率系数}$$
(3.3.5)

$$电池组件的串联数 = \frac{系统工作电压(V) \times 1.43}{组件峰值工作电压(V)} \quad (3.3.6)$$

电池组件(方阵)总功率(W) = 组件并联数 × 组件串联数 × 选定组件的峰值输出功率(W)
(3.3.7)

在进行太阳能电池组件的设计与计算时,还要考虑季节变化对系统发电量的影响。因为在设计和计算得出组件容量时,一般都是以当地太阳能辐射资源的参数,如峰值日照时数、年辐射总量等为参照数据,这些数据都是全年平均数据,参照这些数据计算出的结果,在春、夏、秋季一般都没有问题,冬季可能就会有点欠缺。因此在有条件时或设计比较重要的光伏发电系统时,最好以当地全年每个月的太阳能辐射资源参数分别计算各个月的发电量,其中的最大值就是一年中所需要的电池组件的数量。例如,某地计算出冬季需要的太阳能组件数量是 8 块,但在夏季可能有 5 块就够了,为了保证该系统全年的正常运行,就只好按照冬季的数量确定系统的容量。

(3) 蓄电池和蓄电池组的基本计算中相关因素的考虑

蓄电池和蓄电池组的基本计算公式只是对蓄电池容量的基本估算,在实际应用中还有一些性能参数会对蓄电池的容量和使用寿命产生影响,其中主要的两个因素是蓄电池的放电率和使用环境温度。

① 放电率对蓄电池容量的影响　所谓放电率也就是放电时间和放电电流与蓄电池容量的比率,一般分为 20 小时率 (20h)、10 小时率 (10h)、5 小时率 (5h)、3 小时率 (3h)、1 小时率 (1h)、0.5 小时率 (0.5h) 等。大电流放电时,放电时间短,蓄电池容量会比标称容量缩水;小电流放电,放电时间长,实际放电容量会比标称容量增加。比如,容量 100Ah 的蓄电池用 2A 的电流放电能放 50 个小时,但要用 50A 的电流放电就肯定放不了 2 个小时,实际容量就不够 100Ah 了。蓄电池的容量随着放电率的改变而改变,这样就会对容量设计产生影响。当系统负载放电电流大时,蓄电池的实际容量会比设计容量小,会造成系统供电量不足;而系统负载工作电流小时,蓄电池的实际容量就会比设计容量大,会造成系统成本的无谓增加。特别是在光伏发电系统中应用的蓄电池,放电率一般都较慢,差不多都在 50 小时率以上,而生产厂家提供的蓄电池标称容量是 10 小时率下的容量。因此在设计时要考虑到光伏系统中蓄电池放电率对容量的影响因素,并计算光伏系统的实际平均放电率,根据生产厂家提供的该型号蓄电池在不同放电速率下的容量,就可以对蓄电池的容量进行校对和修正。当手头没有详细的容量-放电速率资料时,也可对慢放电率 50~200h(小时率)光伏系统蓄电池的容量进行估算,一般相对应的比蓄电池的标准容量提高 5%~20%,相应的放电率修正系数为 0.95~0.8。光伏系统的平均放电率计算公式为:

$$平均放电率 = \frac{连续阴雨天数 \times 负载工作时间}{最大放电深度} \quad (3.3.8)$$

对于有多路不同负载的光伏系统,负载工作时间需要用加权平均法进行计算。加权平均

负载工作时间的计算方法为:

$$负载工作时间 = \frac{\sum 负载功率 \times 负载工作时间}{\sum 负载功率} \quad (3.3.9)$$

根据上面两个公式就可以计算出光伏系统的实际平均放电率,根据蓄电池生产厂商提供的该型号蓄电池在不同放电速率下的蓄电池容量,就可以对蓄电池的容量进行修正。

② 环境温度对蓄电池容量的影响　蓄电池的容量会随着蓄电池温度的变化而变化,当蓄电池的温度下降时,蓄电池的容量会下降,温度低于0℃以下时,蓄电池容量会急剧下降。温度升高时,蓄电池容量略有升高。蓄电池温度与放电容量关系曲线如图3.3.1所示。蓄电池的标称容量一般都是在环境温度25℃时标定的,随着温度的降低,0℃时的容量大约下降到标称容量的95%～90%,－10℃时大约下降到标称容量的90%～80%,－20℃时大约下降到标称容量的80%～70%,所以必须考虑蓄电池的使用环境温度对其容量的影响。当最低气温过低时,还要对蓄电池采取相应的保温措施,如地埋、移入房间,或者改用价格更高的胶体铅酸蓄电池等。

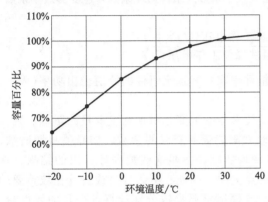

图3.3.1　蓄电池温度与放电容量关系曲线图

当光伏系统安装地点的最低气温很低时,设计时需要的蓄电池容量就要比正常温度范围的容量大,这样才能保证光伏系统在最低气温时也能提供所需的能量。因此,在设计时可参考蓄电池生产厂家提供的蓄电池温度-容量修正曲线图,从该图上可以查到对应温度蓄电池容量的修正系数,将此修正系数纳入计算公式,就可对蓄电池容量的初步计算结果进行修正。如果没有相应的蓄电池温度-容量修正曲线图,也可根据经验确定温度修正系数,一般0℃时修正系数可在0.95～0.9之间选取;－10℃时在0.9～0.8之间选取;－20℃时在0.8～0.7之间选取。

另外,过低的环境气温还会对最大放电深度产生影响。当环境气温在－10℃以下时,浅循环型蓄电池的最大放电深度可由常温时的50%调整为35%～40%,深循环型蓄电池的最大放电深度可由常温时的75%调整到60%。这样既可以提高蓄电池的使用寿命,减少蓄电池系统的维护费用,同时系统成本也不会太高。

(4) 实用的蓄电池容量计算公式

在考虑到各种因素的影响后,将相关系数纳入到上述公式中,才是一个设计和计算蓄电池容量的实用完整公式。即:

$$蓄电池容量 = \frac{负载日平均用电量(Ah) \times 连续阴雨天数 \times 放电率修正系数}{最大放电深度 \times 低温修正系数} \quad (3.3.10)$$

当确定了所需的蓄电池容量后,就要进行蓄电池组的串并联设计。下面介绍蓄电池组串并联组合的计算方法。蓄电池都有标称电压和标称容量,如2V、6V、12V和50Ah、300Ah、1200Ah等。为了达到系统的工作电压,就需要把蓄电池串联起来给系统和负载供电,需要串联的蓄电池个数就是系统的工作电压除以所选蓄电池的标称电压,需要并联的蓄电池数就是蓄电池组的总容量除以所选定蓄电池单体的标称容量。蓄电池单体的标称容量可以有多种选择,例如,假如计算出来的蓄电池容量为600Ah,那么可以选择1个600Ah的单体蓄电池,也可以选择2个300Ah的蓄电池并联,还可以选择3个200Ah或6个100Ah的蓄电池并联。从理论上讲,这些选择都没有问题,但是在实际应用当中,要尽量选择大容量的蓄电池以减少并联的数目。这样做的目的是尽量减少蓄电池之间的不平衡所造成的影响。并联的组数越多，发生蓄电池不平衡的可能性就越大。一般要求并联的蓄电池数量不得超过 4 组。蓄电池串、并联数的计算公式为：

$$蓄电池串联数 = \frac{系统工作电压}{蓄电池标称电压} \quad (3.3.11)$$

$$蓄电池并联数 = \frac{蓄电池总容量}{蓄电池标称容量} \quad (3.3.12)$$

【项目实施】

① 计算　某地建设一个移动通信基站的太阳能光伏供电系统，该系统采用直流负载，负载工作电压 48V，用电量为每天 150Ah。该地区最低的光照辐射是 1 月份，其倾斜面峰值日照时数是 3.5h。选定 125W 太阳能电池组件，其主要参数：峰值功率 125W，峰值工作电压 34.2V，峰值工作电流 3.65A。计算太阳能电池组件使用数量及太阳能电池方阵的组合设计。

太阳能电池组件和蓄电池选择案例

② 计算　某地建设一个移动通信基站的太阳能光伏供电系统，该系统采用直流负载，负载工作电压 48V。该系统有两套设备负载，一套设备工作电流为 1.5A，每天工作 24h；另一套设备工作电流 4.5A，每天工作 12h。该地区的最低气温是 -20℃，最大连续阴雨天数为 6 天，选用深循环型蓄电池，计算蓄电池组的容量和串并联数量及连接方式。

【项目作业】

① 根据项目实施中列出的参数和要求，计算太阳能电池组件的使用数量、太阳能电池方阵的组合方式和串并联示意图。

② 根据项目实施中列出的参数和要求，计算蓄电池组的容量、串并联数量及连接方式示意图。

③ 计算蓄电池的容量和数量的意义有哪些？

项目四　风力机特性仿真

【项目描述】

了解风力机的工作特性，熟悉影响风力机输出功率高低的因素，为提高风力机发电效率做好理论分析的基础。

【能力目标】

① 了解风力机的工作特性。
② 熟悉影响风力机输出功率高低的因素。
③ 了解中国风电产业概况，提高节约资源的意识。

中国风电产业发展概况　　风力发电机介绍

【项目环境】

完成该实训任务需要参考 THWPFG-4 型风光互补发电系统发电系统实训平台设备说明手册，学习模拟风能控制单元电气原理图。

模拟风能控制单元的电路框图如图 3.4.1 所示。

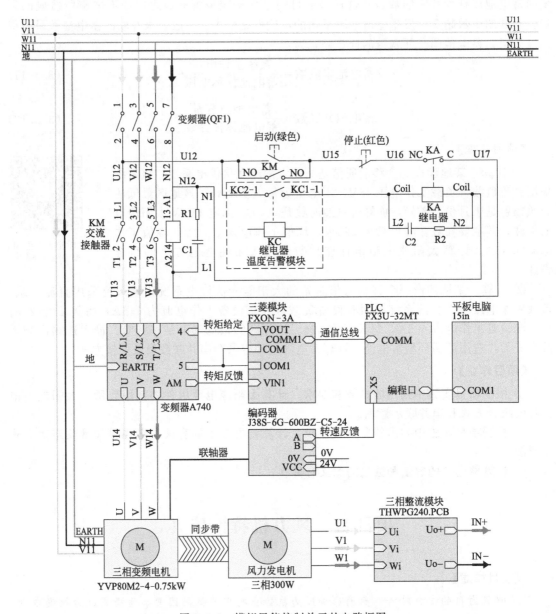

图 3.4.1 模拟风能控制单元的电路框图

模拟风能控制单元主要由模拟风能装置、变频器、交流接触器、电磁继电器、控制开关、连接模块、三相整流模块和阻容吸收模块组成。

① 模拟风能装置 模拟风能装置主要由一台原动机和一台发电机组成，原动机通过同步带带动发电机转动，模拟风力发电的过程。

② 变频器 改变交流电机供电的频率和幅值，达到平滑控制电动机转速的目的。

③ 交流接触器 在模拟风能系统中配合电磁继电器和控制开关控制变频器的启动和停止。

④ 电磁继电器 可以进行信号的隔离、自动调节、安全保护、转换电路等，在模拟风能系统中配合交流接触器及控制开关控制变频器的启动和停止。

⑤ 控制开关　主要完成启动或停止变频器。

⑥ 连接模块　在被控对象与控制器之间起连接作用，方便插拔连接线。

⑦ 三相整流模块　将发电机发出的三相交流电整流为直流电输出。

⑧ 阻容吸收模块　滤除交流接触器和继电器吸合及断开时产生的干扰脉冲。

模拟风能控制单元各组成部分介绍如下。

(1) 模拟风能装置

模拟风能装置的结构如图 3.4.2 所示，主要由原动机、同步带和发电机构成。

原动机技术参数

名称：变频调速三相异步电动机

型号：YVP802-4

额定电压：AC 380V

额定电流：2A

额定功率：0.75kW

图 3.4.2　模拟风能装置

发电机技术参数

名称：MLH 型风力发电机

型号：MLH-300W

额定转速：400r/min

额定功率：300W

输出电压：AC 24V

同步带技术参数

名称：橡胶同步带

型号：424XL

节线长：1076.90mm

齿数：212

图 3.4.3　变频器

(2) 变频器（图 3.4.3）

变频器可以优化电机运行，所以也能起到增效节能的作用。

技术参数

功率范围：0.4～0.75kW

通用矢量控制，1Hz 时 150% 的转矩输出

采用长寿命元器件

内置 modbus-RTU 协议

内置制动晶体管

扩充 PID，三角波功能

带安全停止功能

变频器结构（图 3.4.4）

变频器参数设置

① 操作面板各部分名称（图 3.4.5）

② 基本操作（出厂时设定值，图 3.4.6）

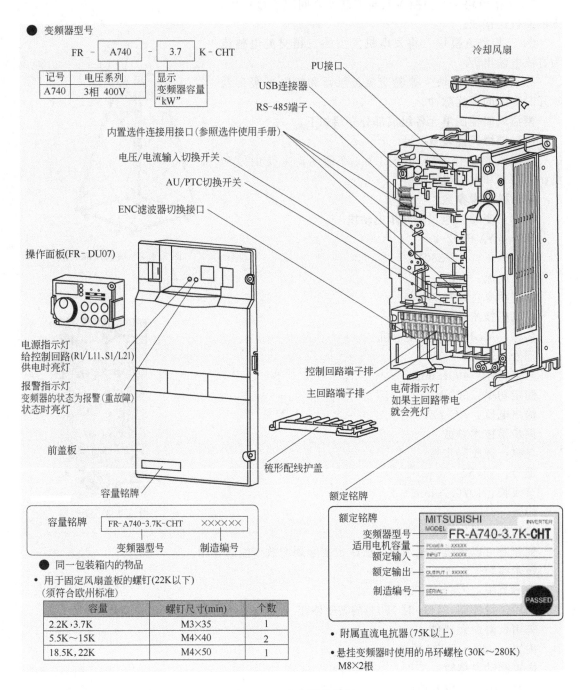

图 3.4.4　变频器结构图

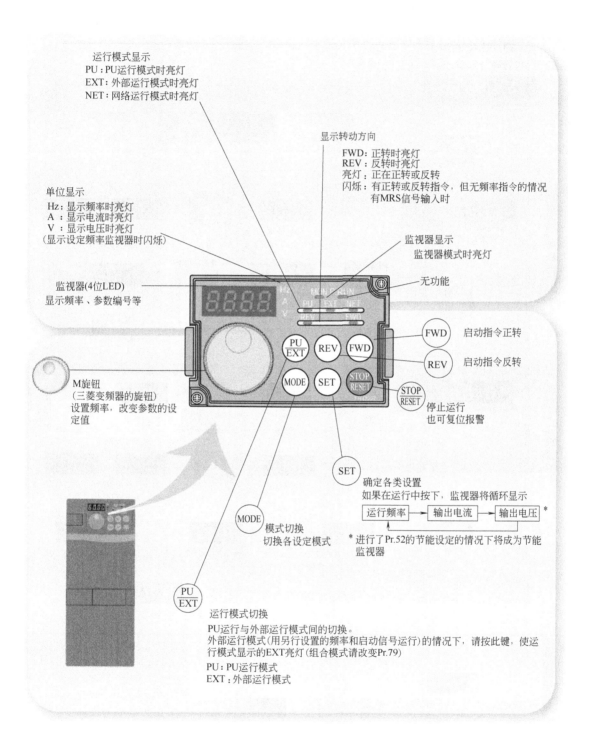

图 3.4.5　变频器的操作面板

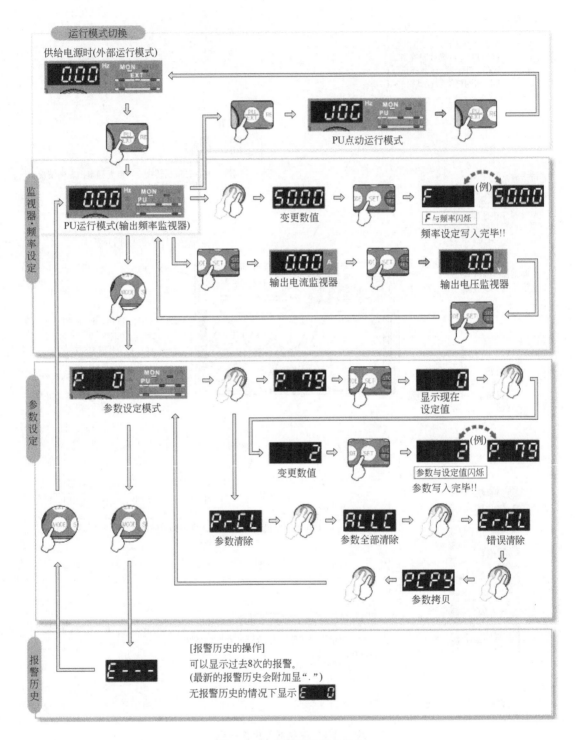

图 3.4.6 变频器的基本操作

(3) 断路器

能接通、分断承载线路的正常电流，也能在规定的异常电路条件下（例如短路）和一定时间内接通、分断承载电流的机械式开关电器。适用于照明配电系统或电动机的配电系统。主要用于交流 50Hz/60Hz，单极 230V，二、三、四极 400V 线路的过载、短路保护，同时也可以在正常情况下不频繁地通断电气装置和照明线路。

各设备开关顺序，如表 3.4.1 所示。

表 3.4.1　各设备开关顺序

序号	功能	风能通电顺序	风能断电顺序	光伏跟踪通电顺序	光伏跟踪断电顺序
1	变频器	1	1	×	×
2	PLC	×	×	2	2
3	开关电源	×	×	1	3
4	模拟光源	×	×	3	1

(4) 交流接触器（图 3.4.7）

技术参数

① 型号　LC1-D1810M5N　220V

② 品牌　施耐德电气-TE 电器

③ 额定电压　220V

④ 额定工作电流　18A

⑤ 额定功率　4kW

⑥ 触点类型　1NC

⑦ 线圈控制电压　AC 220V

⑧ 线圈频率　50Hz

端口定义（表 3.4.2）

图 3.4.7　交流接触器

表 3.4.2　交流接触器各端口定义

序号	定义	说明	序号	定义	说明
1	L1	U 相输入	4	T2	V 相输出
3	L2	V 相输入	6	T3	W 相输出
5	L3	W 相输入	14	NO	KM 2 脚
13	NO	KM 1 脚	A1	A1	线圈 1
2	T1	U 相输出	A2	A2	线圈 2

图 3.4.8　电磁继电器及底座

(5) 电磁继电器及底座（图 3.4.8）

技术参数

① 型号　ARM2F-L

② 线圈电压　AC 220V

③ 触点容量　5A/28V DC/240V AC

④ 安装方式　插拔式（带灯指示）

端口定义（表 3.4.3）

表 3.4.3　电磁继电器及底座端口定义

序号	定义	说明	序号	定义	说明
1	NC	常闭触点	9	C	公共触点
4	NC	常闭触点	12	C	公共触点
5	NO	常开触点	13	Coil	线圈
8	NO	常开触点	14	Coil	线圈

(6) 控制开关（图 3.4.9）

控制变频器的启动和停止。"绿色"按钮为启动，"红色"按钮为停止。

(7) 连接模块（图 3.4.10）

一个连接模块提供两个 10 芯连接器，最多可使用 20 根连接线。

(8) 三相整流模块（图 3.4.11）

将发电机发出的三相交流电整流为直流电输出。

图 3.4.9　控制开关

 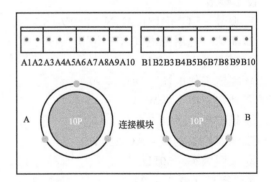

图 3.4.10　连接模块

 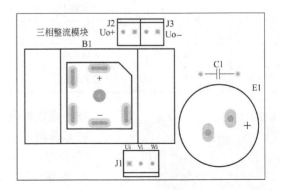

图 3.4.11　三相整流模块

原理图（图 3.4.12）

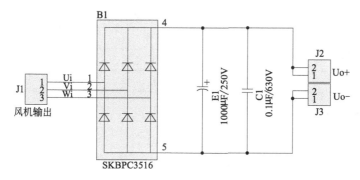

图 3.4.12 三相整流模块原理图

端口定义（表 3.4.4）

表 3.4.4 三相整流模块端口定义

序号	定义	说明	序号	定义	说明
1	J1:Ui	发电机 U 相输出	3	J1:Wi	发电机 W 相输出
2	J1:Vi	发电机 V 相输出	4	J2:Uo+	直流输出

(9) 阻容吸收模块（图 3.4.13）

滤除交流接触器和继电器吸合及断开时产生的干扰脉冲。

 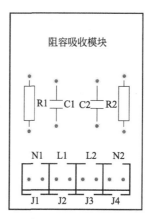

图 3.4.13 阻容吸收模块

原理图（图 3.4.14）

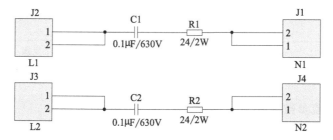

图 3.4.14 阻容吸收模块原理图

端口定义（表3.4.5）

表3.4.5　阻容吸收模块端口定义

序号	定义	说明	序号	定义	说明
1	J1:N1	交流接触器线圈1	3	J3:L2	电磁继电器线圈1
2	J2:L1	交流接触器线圈2	4	J4:N2	电磁继电器线圈2

（10）旋转编码器

型号说明（表3.4.6）

表3.4.6　旋转编码器的型号说明

J38S	-06	-1000	B	Z	-C	5-24
外壳尺寸	输出轴尺寸	分辨率	输出相	零位信号	输出方式	工作电压
ϕ38mm	05=ϕ5mm 06=ϕ6mm 08=ϕ8mm	10~2500P/R	B:两相输出A、B	S=无Z信号 M=有Z信号输出"1" N=有Z信号输出"0"	T=电压型输出 NPN+R C=NPN 集电极开路 P=互补型电路 NPN+PNP L=长线驱动器(26LS31) O=长线驱动器(7272) V=长线驱动器 OC(7273)	5=+5V 526=+5~+26V

电气参数（表3.4.7）

表3.4.7　旋转编码器的电气参数

产品类型	增量式	响应频率	300kHz
电源电压	5V DC,5~26V DC	输出电压	H:VCC×70%，L:0.8V
消耗电流	≤150mA		

机械及环境参数（表3.4.8）

表3.4.8　旋转编码器的机械及环境参数

最大转速	6000r/min	使用温度	-30~+85℃
启动扭矩	0.05N·m	保存温度	-35~+95℃
径向负荷	50N	抗冲击	50g,11ms
轴向负荷	20N	抗震动	10g,10~2000Hz

接线表（表3.4.9）

表3.4.9　旋转编码器的接线表

信号	A	B	Z	A反相	B反相	Z反相	VCC	0V
颜色	绿	白	黄	棕	灰	橙	红	黑

（11）机械及环境参数

① 三菱模块介绍　FX0N-3A 模拟特殊功能块有两个输入通道和一个输出通道，输入通道接收模拟信号并将模拟信号转换成数字值，输出通道采用数字值并输出等量模拟信号。FX0N-3A 的最大分辨率为8位。

② 端子布局和接线　如图 3.4.15 所示。

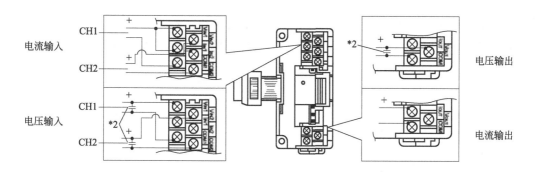

图 3.4.15　FXON-3A 模块的接线

当使用电流输入时，确保标记为【VIN*1】和【IIN*1】的端子连接了。当使用电流输出时，不要连接［VOUT］和［IOUT］端子。*1 此处识别端子编号 1 或 2。

如果电压输入/输出方面出现任何电压波动或者有过多的电噪声，则要在位置*2 连接一个额定值大约在 25V、0.1~0.47μF 的电容器。

【项目原理及基础知识】

由空气动力学可知，风的动能与风速的平方成正比，风的功率与风速的立方成正比。在风速为 v 时，风力机的输入功率 P_{in} 可以用下式表示：

$$P_{in} = \frac{1}{2}\rho A v^3 \tag{3.4.1}$$

$$A = \pi R_w^2 \tag{3.4.2}$$

式中，ρ 为空气密度；A 为风轮扫掠面积；v 为风速；R_w 为风轮半径。

如果通过叶轮扫掠面的风能全部被风机叶片吸收，那么经过叶轮后风速应该等于零，然而空气不可能完全静止不动，因此风力机的效率总小于 1。由此可以定义风力机的风能利用系数 C_p：

$$C_p = \frac{P_o}{P_{in}} \tag{3.4.3}$$

式中，P_o 为风机输出的轴功率。

风机输出的轴功率 P_o 可以用下式表示：

$$P_o = \frac{1}{2}\rho A v^3 C_p \tag{3.4.4}$$

风能利用系数 C_p（也称功率系数）是表征风力机效率的重要参数。C_p 特性一般由风力机制造厂家通过实验给出，本实训中的 C_p 则采用依据统计学公式给出：

$$C_p = 0.5\left(\frac{RC_f}{\lambda} - 0.22\beta - 2\right)e^{-0.255\frac{RC_f}{\lambda}} \tag{3.4.5}$$

式中，C_f 为叶片设计常数，一般取 1~3；桨距角 β 是来流速度方向与弦线间的夹角。

影响 C_p 的因素有风速、叶片转速、风轮半径、桨距角，因此风力机 C_p 特性比较复杂，但是风力机的最大风能利用系数 C_{pmax} 通过"贝茨极限理论"可知，$C_{pmax} \approx 0.593$。

为了便于分析 C_p 特性，定义风力机的另一个重要参数叶尖速比 λ，即风力机叶片尖端线速度与风速之比：

$$\lambda = \frac{R_w \omega_w}{v} = \frac{\pi n_w}{30v} \tag{3.4.6}$$

式中，ω_w 为叶片旋转角速度，rad/s；n_w 为叶片转速，r/min。

风力机特性通常用一簇风能利用系数 C_p 的曲线来表示，如图 3.4.16 和图 3.4.17 所示。

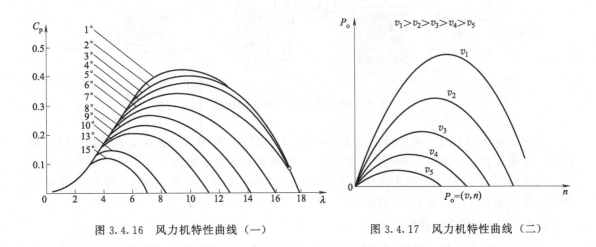

图 3.4.16 风力机特性曲线（一）　　　　图 3.4.17 风力机特性曲线（二）

由图 3.4.16 可以看出，$C_p(\lambda)$ 曲线是桨距角 β 的函数，且当 β 角增大时，$C_p(\lambda)$ 曲线显著缩小。

【项目实施】

(1) 风能利用系数与速比的曲线观测

① 双击桌面"独立服务器"图标，进入监控系统后，点击"登录"，然后在顶窗口点击"风力机特性仿真"按钮，进入仿真界面。将实时风速设定为 6m/s，其余参数用默认值，依次点击"自动绘制曲线""启动"，在 C_p/λ 图中可以看出 C_p 随 λ 的变化曲线。在 P_1/n、P_0/n 图中可观测风机

风力机特性仿真

的输入功率、输出功率、转矩随角速度的变化曲线。完成后点击"停止"。

② 将实时风速给定为 12m/s，重复上述步骤，观测其 C_p/λ、P_1/n、P_0/n 曲线与 6m/s 的区别。

(2) 桨距角对风能利用系数的影响

① 将桨距角设置为 5°，重复上述步骤，观测其 C_p/λ、P_1/n、P_0/n 曲线与 0°时的区别。

② 将桨距角设置为 20°，重复上述步骤，观测其 C_p/λ、P_1/n、P_0/n 曲线与 0°、5°的区别。

(3) 手动点绘 C_p/λ、P_1/n、P_0/n 曲线

进入风电监控软件，点击"风力机特性仿真"，进入仿真界面，实时风速设定为 6m/s，其余参数用默认值。依次点击"手动绘制曲线""启动"后，在风轮转速设定值栏输入风轮转速后点"确认"，点击"记录"，在 C_p/λ、P_1/n、P_0/n 图中可观测到与风轮转速对应的点。在风轮转速显示数值软键盘中输入不同风轮转速，重复点击"记录"，可手动绘出 C_p/λ、P_1/n、P_0/n 曲线，完成后点击"停止"。

【项目作业】
① 记录桨距角固定、不同风速时手动点绘的 C_p/λ、P_1/n、P_0/n 曲线。
② 记录风速固定、不同桨距角时手动点绘的 C_p/λ、P_1/n、P_0/n 曲线。
③ 综合上述曲线,分析风力机的工作特性。
④ 请描述模拟风能装置的主要结构。

项目五 验证光伏阵列最大功率跟踪算法

【项目描述】
掌握 MPPT 算法编程,了解能源转换储存系统的软件及硬件组成、软件的基本操作及数据采集操作,并能依据控制信号及采集数据对本系统的工程项目进行二次开发。

【能力目标】
① 了解光伏阵列最大功率跟踪算法。
② 掌握光伏阵列最大功率跟踪算法的原理及应用。

【项目环境】
完成该实训任务需要参考 THWPFG-4 型风光互补发电系统发电系统实训平台设备说明手册,学习能源转换储存控制单元电气原理图,了解 MPPT 控制器组成、电气指标及接口描述、MPPT 控制器使用原理。了解 CPU 核心模块工作原理,根据光伏阵列电压、电流采样信号进行最大功率跟踪的程序设计和调试。编写不同的 MPPT 算法实现最大功率跟踪,并将调节参数(占空比参数)通过串口发送给 PWM 驱动模块进行调节。微处理器采用 51 系列单片机,具有在线下载功能,方便编程调试,实现 MPPT 控制算法。

光伏阵列最大功率跟踪调试

(1) MPPT 控制器的组成

MPPT 控制器主要由直流电压电流采集模块 1、直流电压电流采集模块 2、CPU 核心模块、人机交互模块、PWM 驱动模块、通信接口模块、温度告警模块、DC-DC Boost 主电路模块、智能充放电控制器组成,完成对太阳能电池的最大功率跟踪,有效地提高太阳能电池的工作效率,同时也改善系统的工作性能。MPPT 控制器功能框图如图 3.5.1 所示。

MPPT 控制器各功能模块说明如下。

① 直流电压电流采集模块 1 通过电压霍尔传感器和电流传感器将光伏阵列电池输出的电压和电流转换成满足单片机输入端要求的电压信号。

② 人机交互模块 CPU 核心模块的输入、输出终端。

③ 温度告警模块 检测蓄电池的温度,当超过设定的参考温度时,液晶显示电池温度过高。

④ PWM 驱动模块 接收最大功率跟踪微处理器输出的功率调节参数(占空比参数),并输出不同占空比的 PWM 信号,将 PWM 微处理输出的 PWM 信号与 DC/DC 电路隔离,并将 PWM 信号转换成满足开关管需求的驱动信号,提高驱动能力。

⑤ 直流电压电流采集模块 2 通过电压霍尔传感器和电流传感器,将输出的电压和电流

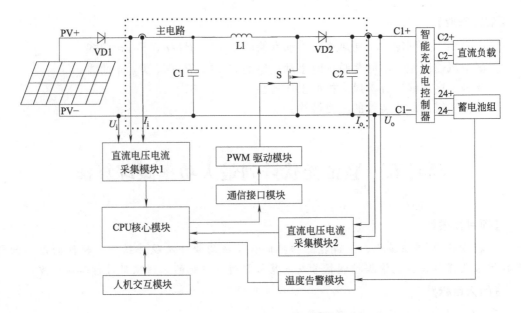

图 3.5.1　MPPT 控制器功能框图

转换成满足单片机输入端要求的电压信号。

⑥ 通信接口模块　主要起信号电平转换作用，将 TTL 电平转换成 RS-232 和 RS-485 信号。

⑦ DC-DC Boost 主电路模块　通过实时采集光伏阵列电池的功率来调整主电路的占空比，等效调节了负载的阻抗，使负载取用的功率（即伏/秒面积）得以改变，可始终跟随光伏阵列电池输出的最大功率点，等效为一个阻抗变换器。

⑧ 智能充放电控制器　根据蓄电池电压高低，调节充电状态和电流的大小，防止蓄电池过充或过放，延长蓄电池使用寿命。

(2) MPPT 控制器的 CPU 核心模块

图 3.5.2 为 CPU 核心模块实物图，表 3.5.1 为 CPU 核心模块端口定义。

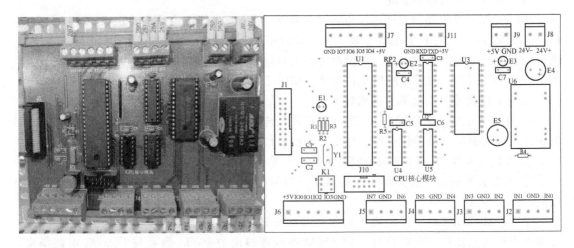

图 3.5.2　CPU 核心模块实物图

表 3.5.1　CPU 核心模块端口定义

序号	名称	说明	扩展接口	备注
1	J2:IN0	AD 采样输入 0,5V 以内电压信号		
2	J2:GND			
3	J2:GND	AD 采样输入 1,5V 以内电压信号		
4	J2:IN1			
5	J3:IN2	AD 采样输入 2,5V 以内电压信号		
6	J3:GND			
7	J3:GND	AD 采样输入 3,5V 以内电压信号		
8	J3:IN3			
9	J4:IN4	AD 采样输入 4,5V 以内电压信号		
10	J4:GND			
11	J4:GND	AD 采样输入 5,5V 以内电压信号		
12	J4:IN5			
13	J5:IN6	AD 采样输入 6,5V 以内电压信号	√	
14	J5:GND		√	
15	J5:GND	AD 采样输入 7,5V 以内电压信号	√	
16	J5:IN7		√	
17	J6:+5V	+5V 电源输出	√	
18	J6:IO0	开关量输入输出接口 0,5V 以内电压信号	√	
19	J6:IO1	开关量输入输出接口 1,5V 以内电压信号	√	
20	J6:IO2	开关量输入输出接口 2,5V 以内电压信号	√	
21	J6:IO3	开关量输入输出接口 3,5V 以内电压信号	√	
22	J6:GND	地	√	
23	J7:+5V	+5V 电源输出	√	
24	J7:IO4	开关量输入输出接口 4,5V 以内电压信号	√	
25	J7:IO5	开关量输入输出接口 5,5V 以内电压信号	√	
26	J7:IO6	开关量输入输出接口 6,5V 以内电压信号	√	
27	J7:IO7	开关量输入输出接口 7,5V 以内电压信号	√	
28	J7:GND	地	√	
29	J11:+5V	+5V 电源输出		
30	J11:TXD	串行口发送端		
31	J11:RXD	串行口接收端		
32	J11:GND	地		
33	J9:+5V	隔离电源 DC/DC 输出 5V 电源正极	√	
34	J9:GND	隔离电源 DC/DC 输出 5V 电源负极	√	
35	J8:24V+	隔离电源 DC/DC 输入 24V 正极		
36	J8:24V−	隔离电源 DC/DC 输入 24V 负极		
37	J1	至人机交互模块		20 排线

编程注意事项

① MPPT 控制器内含有智能充放电控制器。由于智能充放电控制器在上电之后需要至少 1min 才能完成启动工作，因此在设计最大功率跟踪程序时需要延时 1min，以确保智能充放电控制器正常工作。

② CPU 核心模块微处理器与 PWM 驱动模块微处理器之间的通信采用串口通信，波特率 9600；8 位数据位；1 位停止位；无校验位；1 个 8Bit 表示占空比，数据范围 0x00～0xFF（0x00：占空比为 0%；0xFF：占空比为 99%）。

③ 四路待转换的模拟信号分别接在 ADC0809 的 IN0、IN1、IN2、IN3 输入通道上，汇编采用累加器 A 与外部数据存储器传送指令选择通道和读取 AD 值。

各通道的访问地址如下所示：
光伏阵列电池电流（IN0）地址　0x7FF8；
光伏阵列电池电压（IN1）地址　0x7FF9；
Boost 变换器输出电流（IN2）地址　0x7FFA；
Boost 变换器输出电压（IN3）地址　0x7FFB。
选择通道用汇编指令：
MOVX @DPTR，A（其中 DPTR 为地址，A 可以为任何值【0x00～0xFF】）
读取 AD 值用汇编指令：
MOVX A，@DPTR（其中 DPTR 为地址，A 为 AD 转换值）

④ 四路待转换的模拟信号分别接在 ADC0809 的 IN0、IN1、IN2、IN3 输入通道上，采用 C 语言需要预定义各通道为外部数据区，定义如下：

```
#define  II_CURRENT    XBYTE[0x7FF8]    /*光伏阵列电池电流*/
#define  UI_VOL        XBYTE[0x7FF9]    /*光伏阵列电池电压*/
#define  IO_CURRENT    XBYTE[0x7FFA]    /* Boost 变换器输出电流*/
#define  UO_VOL        XBYTE[0x7FFB]    /* Boost 变换器输出电压*/
```

选择通道用 C 语句：
XXXX=X（其中 XXXX 为预定义名，X 可以为任意值【0x00～0xFF】）
读取 AD 值用 C 语句：
变量=XXXX（其中 XXXX 为预定义名，变量的内容为 AD 转换值）

⑤ 编程示例（电压回授法主程序示例）

```
UART_Init();                          //串口初始化
duty= DUTY_DEFAULT;                   //设置默认占空比参数
uart_send_duty(duty);                 //发送占空比参数给 PWM 微处理器
DelayMS(50);                          //延时
average_ui=average(0);                //采集开路电压值
v_const=(average_ui* 3)>> 0x02;       //计算参考电压 V_const=0.75*Voc
while(1)
{
  average_ui=average(0);              //输入电压平均值
  v_error=cabs(v_const - average_ui); //计算参考电压与输入电压的绝对误差值
  if(v_error>ERROR_VOLTAGE)           //绝对误差值大于固定值,调整占空比,否则占空比不变
  {
    if(v_const>average_ui)            //参考电压大于输入电压,减小输出电压,增大输入电压
    {
```

```
      duty=duty-DUTY_STEP;          //减小占空比,减小输出电压
    }
    else                             //参考电压小于输入电压,增大输出电压,减小输入电压
    {
      duty=duty+DUTY_STEP;          //增大占空比,增大输出电压
    }
    uart_send_duty(duty);           //发送占空比参数给PWM微处理器
    DelayMS(50);                    //延时
  }
}
```

(3) MPPT 控制器的人机交互模块

图 3.5.3 为人机交互模块实物照片，表 3.5.2 为人机交互模块端口定义。

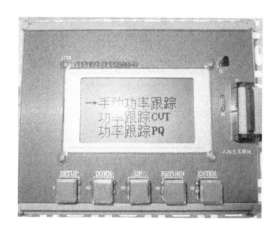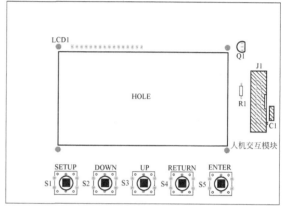

图 3.5.3　人机交互模块实物照片

表 3.5.2　人机交互模块端口定义

序号	名称	说明	序号	名称	说明
1	ENTER	确定按键	11	R/W	读/写信号
2	RETURN	返回按键	12	D7	数据位 7
3	INT1	按键公共端	13	E	使能信号
4	UP	向上按键	14	D6	数据位 6
5	GND	地	15	D0	数据位 0
6	DOWN	向下按键	16	D5	数据位 5
7	VCC	电源	17	D1	数据位 1
8	SETUP	设置按键	18	D4	数据位 4
9	RS	寄存器选择	19	D2	数据位 2
10	P10	背光控制	20	D3	数据位 3

【项目原理及基础知识】

(1) 电压回授法

早期对光伏电池输出功率控制主要利用电压回授（Constant Voltage Tracking，CVT）技术。图 3.5.4 中，L 是负载特性曲线，当温度保持某一固定值时，在不同的日照强度下与伏安特性曲线的交点 a、b、c、d、e 对应于不同的工作点。人们发现阵列可提供最大功率

的那些点,如 a'、b'、c'、d'、e' 点连起来,几乎落在同一根垂直线的邻近两侧,这就有可能把最大功率点的轨迹线近似地看成电压 $U=$ 常数的一根垂直线,亦即只要保持阵列的输出端电压 U 为常数,就可以大致保证阵列输出在该温度下的最大功率,于是最大功率点跟踪器可简化为一个稳压器。这种方法实际上是一种近似最大功率法。

CVT 控制方式具有控制简单、可靠性高、稳定性好、易于实现等优点,比一般光伏系统可望多获得 20% 的电能。但该跟踪方式忽略了温度对太阳能电池开路电压的影响。以单晶硅电池为例,当环境温度每升高 10℃ 时,其开路电压的下降率为 0.35%~0.45%,这表明光伏电池最大功率点对应的电压 U_n 也随环境温度的变化而变化。对于四季温差或日温差较大地区,CVT 控制方式并不能在所有的温度环境下完全地跟踪最大功率。系统控制流程图如图 3.5.5 所示。

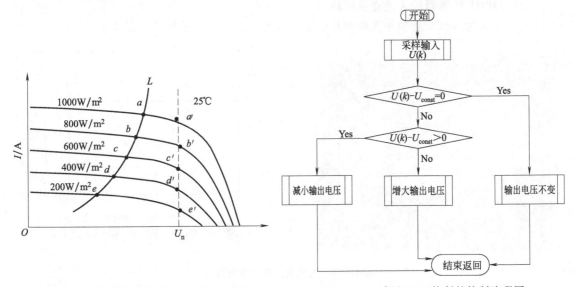

图 3.5.4 硅光伏电池阵列具有的伏安特性　　图 3.5.5 采用 CVT 控制的控制流程图

(2) 扰动观察法

扰动观察法(Perturbation and Observation, P&O)也称为爬山法(Hill Climbing, HC)。其工作原理为测量当前阵列输出功率,然后在原输出电压上增加一个小电压分量扰动后,其输出功率会发生改变,测量出改变后的功率,与改变前的功率进行比较,即可获知功率变化的方向。如果功率增大,就继续使用原扰动,如果功率减小,则改变原扰动方向。扰动观察法跟踪情况如图 3.5.6 所示。

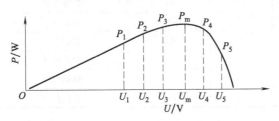

图 3.5.6 扰动观察法跟踪情况示意

假设工作点在 U_1 处,光伏电池输出功率为 P_1,如果使工作点移到 $U_2=U_1+\Delta U$,光伏电池输出功率为 P_2,比较现时功率 P_2 与记忆功率 P_1,若 $P_2>P_1$,说明输入信号差 ΔU 使输出功率变大,工作点位于最大功率值 P_m 的左边,需要继续增大电压,使工作点继续朝右边即 P_m 的方向移动。如果工作点已越过 P_m 到达 U_4,此时若再增加 ΔU,则工作点到达 U_5,比较结果为 $P_5<P_4$,说明工作点位于最大功率值 P_m 的右边,则需要改变输入信号的变化方向,即输入信号每次减去 ΔU 后,再比较现时功率与记忆功率,直至找到最大功率点 P_m。

由于扰动观察法采用模块化控制回路，其结构简单、测量参数少、容易实现，因此广泛应用于光伏电池的最大功率点跟踪。其缺点是到达最大功率点附近后，会在其左右振荡，造成能量损耗，尤其在气候条件变化缓慢时，情况更为严重。因为气候条件变化缓慢时，光伏电池所产生的电压及电流变动并没有什么太大的变化，而此方法仍然会继续扰动以改变其电压值而造成能量损失，虽然可以缩小每次扰动的幅度，以降低 P_m 点的振荡幅度来减少能量损失，不过当温度或光照度有大幅变化时，这种方法会使跟踪到另一个最大功率点的速度变慢，此时会浪费大量能量。因此跟踪步长与跟踪精度和响应速度无法兼顾，有时在运行中会发生程序"误判"现象。系统控制流程图如图3.5.7所示。

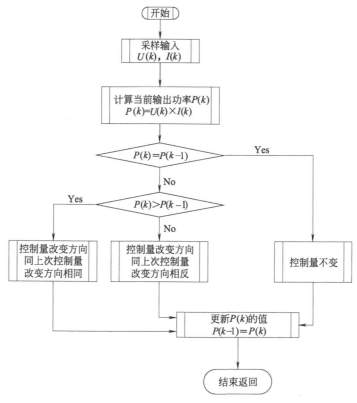

图 3.5.7　扰动观测法控制流程图

编程注意事项和编程示例前面已介绍过（见44页）。

【项目实施】

① 合上"模拟能源控制系统"的"总电源"开关，系统得电，三相电源指示灯亮。

② 合上"开关电源"空气开关，使开关电源工作。

③ 合上"模拟光源"空气开关，使模拟太阳光灯打开。

④ 合上"PLC"空气开关，PLC上电，将PLC上的拨动开关置"RUN"状态，使PLC程序处于运行状态。

⑤ 按下追日系统控制开关的"开始"按钮（绿色按钮），各步进电机准备工作；按下"控制"按钮（黄色按钮），模拟光源开始运动。

⑥ 合上"能源转换储存控制系统"的"总电源"开关，系统得电，三相电源指示灯亮。

光伏阵列最大功率跟踪操作演示

⑦ 记录光伏输出电压，此值即为光伏开路电压。

⑧ 将充放电控制器的刹车置于"RELEASE（自动刹车）"状态，合上"能源转换储存控制系统"的"蓄电池"空气开关，蓄电池接入光伏 MPPT 控制系统，同时给充放电控制器供电，此时充放电控制器进行初始化，红色指示灯点亮（工作在刹车状态）。必须等红色指示灯熄灭（退出刹车状态）才能进行下一步操作。

⑨ 合上"能源转换储存控制系统"的"光伏输出"和"光伏 MPPT"空气开关。

⑩ 按"CPU 核心模块"上的复位按钮 K1，系统复位。

⑪ 按"人机交互模块"的"ENTER"键，进入手动功率跟踪界面，然后按"UP"和"DOWN"键手动调节占空比，测量多组光伏输出电流、光伏输出电压，记录下数值。

⑫ 通过记录的电压、电流数据，计算每个电压、电流对应的功率。

⑬ 通过"人机交互模块"上的"UP"和"DOWN"键，选择"功率跟踪 CVT"或"功率跟踪 PQ"，按"ENTER"键进入自动功率跟踪方式，经过一段时间之后，分别记录电压、电流数据，并计算对应的功率。

⑭ 实验结束后，按下"停止"按钮（红色按钮），所有步进电机停止运行。再依次关闭模拟能源控制系统的"PLC""开关电源""模拟光源"空气开关和总电源开关，能源转换储存控制系统的"光伏 MPPT"、"蓄电池"空气开关和总电源开关。如需进行后续实验，可不关闭总电源。

【项目作业】

① 根据测量的数据，绘制 U-I 曲线、功率曲线，并找出最大功率点。

② 计算哪个点是最大功率输出点，为什么？

③ 比较自动功率跟踪"CVT"和"PQ"两种方式下的最大功率点和手动跟踪方式下的最大功率点，并说明两种自动功率跟踪方式的优缺点。

④ 尝试改编程序试运行，理解最大功率跟踪算法。

项目六　学习离网型逆变器原理并完成系统测试

【项目描述】

学习并网逆变控制单元电路框图、断路器开关顺序、离网逆变器的安装使用方法。

【能力目标】

① 掌握离网型逆变器的作用及使用方法。

② 掌握离网型逆变器的工作原理。

【项目环境】

完成该实训任务需要参考 THWPEG-4 型风光互补发电系统发电系统实训平台设备说明手册。

并网逆变控制单元的电路框图如图 3.6.1 所示。

(1) 断路器

断路器开关顺序如表 3.6.1 所示。

光伏离网逆变器实训

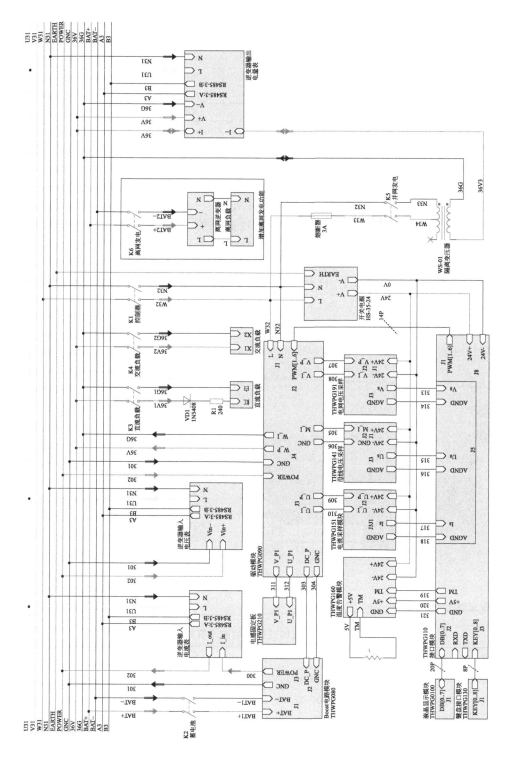

图 3.6.1 并网逆变控制单元电路框图

表 3.6.1 断路器开关顺序

序号	功能	直流电机通电顺序	直流电机断电顺序	交流负载通电顺序	交流负载断电顺序	并网发电通电顺序	并网发电断电顺序	离网发电通电顺序	离网发电断电顺序
K1	控制器	1	3	1	3	1	3	1	3
K2	蓄电池	2	2	2	2	2	2	2	2
K3	直流负载	3	1	×	×	×	×	×	×
K4	交流负载	×	×	3	1	×	×	×	×
K5	并网发电	×	×	×	×	3	1	×	×
K6	离网发电	×	×	×	×	×	×	3	1

(2) 离网逆变器（图 3.6.2）

① 产品特点

a. 使用先进的双 CPU 单片机智能控制技术，具有高可靠性、低故障率的特点。

b. 纯正弦波输出，带负载能力强，应用范围广。

c. 具有完善的保护功能（过负载保护、内部过温保护、输出短路保护、输入欠压保护、输入过压保护等），大大提高产品的可靠性。

d. 内部采用 CPU 集中控制、贴片技术，使得体积非常小、重量轻。

e. 散热风机智能控制，采用 CPU 控制散热风机的工作状态，大大延长风机的使用寿命；并且节约电能，提高工作效率。

f. 工作噪声小，效率高。

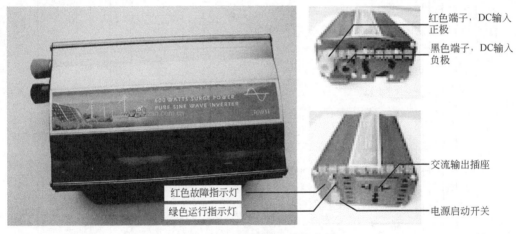

图 3.6.2 离网逆变器实物图

② 技术指标（表 3.6.2）

表 3.6.2 离网逆变器的技术指标表

	型号	MZ-300W
输入	直流电压	12V,24V,48V,110V
	直流电压范围	10～15V DC,21～30V DC,42～60V DC,100～120V DC
	空载电流	<0.5mA
	效率	>85%
	直流连接	带插口或汽车适配器的电缆

续表

输出	输出电压	100/110/120V AC,220/230/240V AC
	额定功率	300W
	瞬间功率	600W
	输出波形	正弦波
	输出频率	50Hz,60Hz
	波形失真	3%
保护	低压报警	10V DC±0.5V,20.5V DC±1V,44V DC±1V,100V DC±1V
	低压关断	9.5V DC±0.5V,19.5V DC±1V,42V DC±1V,96V DC±5V
	过载	输出关闭
	过压	15.5V,30.5V,61.2V,120V
	过热	输出自动关闭
	保险丝	短路
环境	工作温度	−10~+50℃
	湿度	20%~90%RH 无冷凝
	储存温度	−30~+70℃
包装	机器尺寸/mm	180×105×60
	含包装尺寸/mm	245×122×70
	净重/kg	0.78
	毛重/kg	1
	包装方式	彩盒包装
其他	启动	软启动
	冷却方式	风扇冷却
	总波形失真	THD＜5%

③ 安装与使用方法

a.电源选择。必须使用蓄电池、车载电池或者太阳能发电系统电源，使输入的电压为12V/24V/36V/48V/110V，适宜该产品。逆变电源严禁使用高于额定输入电压电源。

b.使开关处于OFF（关）位置（包括逆变电源和设备），从电池处获得电源，使逆变电源与输入电源相连，黑电线接黑端子为负极，红电线接红端子为正极。

c.逆变电源与电器相连。确保逆变电源的负载功率在规定功率内，在打开电源时不应超过逆变电源的最大功率。当逆变电源与设备相连时，打开逆变电源和设备上的开关。开机时红灯反复亮暗是正常的，如超过1min视为不正常，应检查负载是否过大。

④ 保护功能

a.输入欠压保护　当电池电压很低时，报警器发出警报，表明直流电供应电压降低，电池需要再充电。当12V的输入电压低于10V±0.5V，24V的低于20V±1.0V，36V的低于30V±1.0V，48V的低于40V±2.0V，110V的低于90V±2.0V时，交流电输出会自动关闭，指示灯和报警器会同时动作。

b.输入过压保护　对于12V的逆变电源来说，当输入的电压达到15.5V±0.5V，24V的达到31V±1.0V，36V的达到46.5V±1.0V，48V的达到62V±2.0V，110V的达到132V±2.0V时，指示灯变红，同时交流电输出会自动关闭。

c. 短路保护　当发生短路时，输出设备将关闭。
d. 超载保护　当发生超载时，输出设备将关闭，红灯亮。
e. 极性相反输入保护　当电池被相反连接时，保险片会烧掉以此来保护设备，同时也有存在损坏逆变电源内部器件的可能。
f. 过热保护　当内在温度超过75℃时，交流电输出会自动关闭，红色指示灯亮。直到温度降低到正常后，才能够使用。
g. 风扇智能控制　当散热器内部温度超过50℃左右时，内置风扇会自动开启，以使逆变电源冷却。

⑤ 常见故障排除
a. 逆变电源没有反应　检查电池与逆变电源的接触是否不良。重新连接。
b. 输出电压过低。
c. 超载　负载功率超过额定功率。关闭部分设备再重新开启逆变电源。
d. 输入电压过低　确保输入电压在额定范围内。
e. 低压报警。
f. 电池没电　重新换电池。
g. 电池电压过低或接触不良　重新换电池，检查连接，或者用干布清洗连接端子。
h. 逆变电源没有输出。
i. 电池电压过低　重换电池。
j. 负载容量过大　关闭部分设备，重启逆变电源。
k. 逆变电源过热保护　冷却逆变电源并放在通风的地方。
l. 逆变电源开启失败　重新开启逆变电源。
m. 正和负相反连接　保险片烧掉。重换同规格的保险片再重新连接。
n. 逆变电源不运行　检查电源开关，保险片和电池连接是否完好。

【项目原理及基础知识】

通常，把将交流电能变换成直流电能的过程称为整流，把完成整流功能的电路称为整流电路，把实现整流过程的装置称为整流设备或整流器。与之相对应，把将直流电能变换成交流电能的过程称为逆变，把完成逆变功能的电路称为逆变电路，把实现逆变过程的装置称为逆变设备或逆变器。

现代逆变技术是研究逆变电路理论和应用的一门科学技术，是建立在工业电子技术、半导体器件技术、现代控制技术、现代电力电子技术、半导体变流技术、脉宽调制（PWM）技术等学科基础之上的一门实用技术。它主要包括半导体功率集成器件及其应用、逆变电路和逆变控制技术三大部分。

(1) 逆变器的分类

逆变器的种类很多，可按照不同的方法进行分类。

① 按逆变器输出交流电能的频率分　可分为工频逆变器、中频逆变器和高频逆变器。工频逆变器的频率为50~60Hz的逆变器；中频逆变器的频率一般为400Hz到十几千赫；高频逆变器的频率一般为十几千赫到兆赫。

② 按逆变器输出的相数分　可分为单相逆变器、三相逆变器和多相逆变器。

③ 按逆变器输出电能的去向分　可分为有源逆变器和无源逆变器。凡将逆变器输出的电能向工业电网输送的逆变器，称为有源逆变器；凡将逆变器输出的电能输向某种用电负载的逆变器，称为无源逆变器。

④ 按逆变器主电路的形式分 可分为单端式逆变器、推挽式逆变器、半桥式逆变器和全桥式逆变器。

⑤ 按逆变器主开关器件的类型分 可分为晶闸管逆变器、晶体管逆变器、场效应管逆变器和绝缘栅双极晶体管（IGBT）逆变器等。又可将其归纳为"半控型"逆变器和"全控制"逆变器两大类。前者不具备自关断能力，元器件在导通后即失去控制作用，故称之为"半控型"，普通晶闸管即属于这一类。后者则具有自关断能力，即元器件的导通和关断均可由控制极加以控制，故称之为"全控型"，电力场效应晶体管和绝缘栅双极晶体管（IGBT）等均属于这一类。

⑥ 按直流电源分 可分为电压源型逆变器（VSI）和电流源型逆变器（CSI）。前者直流电压近于恒定，输出电压为交变方波；后者直流电流近于恒定，输出电流为交变方波。

⑦ 按逆变器输出电压或电流的波形分 可分为正弦波输出逆变器和非正弦波输出逆变器。

⑧ 按逆变器控制方式分 可分为调频式（PFM）逆变器和调脉宽式（PWM）逆变器。

⑨ 按逆变器开关电路工作方式分 可分为谐振式逆变器、定频硬开关式逆变器和定频软开关式逆变器。

⑩ 按逆变器换流方式分 可分为负载换流式逆变器和自换流式逆变器。

（2）逆变器的基本结构

逆变器的直接功能是将直流电能变换成为交流电能。逆变装置的核心是逆变开关电路，简称为逆变电路。该电路通过电力电子开关的导通与关断，来完成逆变的功能。电力电子开关器件的通断需要一定的驱动脉冲，这些脉冲可通过改变一个电压信号来调节。产生和调节脉冲的电路，通常称为控制电路或控制回路。逆变装置的基本结构，除上述的逆变电路和控制电路外，还有保护电路、输出电路、输入电路、输出电路等。

（3）逆变器的工作原理

单相电压全控型 PWM 逆变器工作原理如图 3.6.3 所示，为通常使用的单相输出的全桥逆变主电路，图中，交流元件采用 IGBT 管 Q11、Q12、Q13、Q14，并由 PWM 脉宽调制控制 IGBT 管的导通或截止。

当逆变器电路接上直流电源后，先由 Q11、Q14 导通，Q12、Q13 截止，则电流由直流电源正极输出，经 Q11、L_o、负载、变压器初级线圈，经过 Q14 回到电源负极。当 Q11、Q14 截止后，Q12、Q13 导通，电流从电源正极输出，经 Q13、变压器初级线圈、负载、L_o，经过 Q12 回到电源负极。此时，在变压器初级线圈上，已形成正负交变方波，利用高频 PWM 控制，两对 IGBT 管交替导通或截止，在变压器上产生交流电压。由于 LC 交流滤波器的作用，使输出端形成正弦波交流电压。

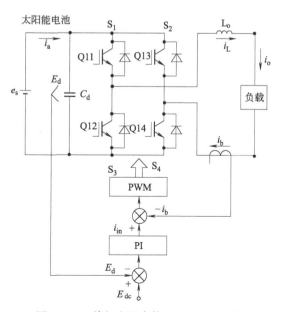

图 3.6.3 单相电压全控型 PWM 逆变器

离网型逆变器工作操作演示

【项目实施】

① 打开各控制系统总电源开关,电源指示灯有显示。

② 将充放电控制器的刹车置于"RELEASE(自动刹车)"状态,合上"能源转换储存控制系统"的"蓄电池"空气开关,给充放电控制器供电。此时充放电控制器进行初始化,红色指示灯点亮(工作在刹车状态)。必须等红色指示灯熄灭(退出刹车状态)才能进行下一步操作。

③ 依次打开"并网逆变控制系统"的"蓄电池"和"离网发电"开关。

④ 启动离网逆变器(离网逆变器安装要反面),观察离网负载(风扇)是否能够正常运转,记录离网逆变器工作前后蓄电池电压表、电流表的值(表3.6.3)。

表 3.6.3 数据记录

序号	项目	蓄电池电压表 U/V	蓄电池电流表 I/A
1	工作前		
2	工作后		

⑤ 实验结束后,依次关闭"并网逆变控制系统"的"离网发电""蓄电池"空气开关和"能源转换储存控制系统"的"蓄电池"空气开关,最后再关闭各控制系统的总电源开关。如需进行后续实验,可不关闭控制系统总电源。

【项目作业】

① 参阅相关资料,简述离网型逆变电源的用处。

② 单相电压型PWM逆变器,Q11、Q14导通,则电流由哪极输出?并经由哪个开关回到电源负极?

项目七 测试并网型逆变器工作原理并分析电能质量

【项目描述】

学习并网逆变控制单元中并网逆变器的组成和并网逆变器的参数设置。

【能力目标】

① 了解并网逆变器的工作原理。

② 掌握并网逆变器的使用及其作用。

③ 了解电能质量及谐波治理。

【项目环境】

(1)并网逆变器的组成

并网逆变器功能框图如图3.7.1所示,主电路拓扑结构由DC/DC(Boost电路模块)+DC/AC(驱动电路模块)+滤波器(滤波板)组成,控制回路由母线电压采样模块+电流

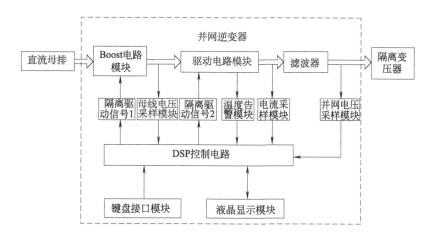

图 3.7.1 并网逆变器功能框图

采样模块＋并网电压采样模块＋温度告警模块＋隔离驱动信号＋DSP 控制电路＋键盘接口模块＋液晶显示模块组成。

① 并网逆变器模块说明

a. Boost 电路模块　Boost 升压电路主要将直流母排输出直流电压变换成能满足并网要求的母线电压。

b. 驱动电路模块　驱动电路将该直流母线电压经过 DC/AC 逆变成与电网电压同频、同相、同幅的正弦交流电，以实现与电网的并网连接。

c. 滤波器（滤波板）　滤除逆变器输出高频 PWM 谐波电流，减小进网电流中的高频环流，又能在逆变器与电网间进行能量的传递，使并网逆变器获得一定的阻尼特性，减小冲击电流，有利于系统的稳定运行。

d. 母线电压采样模块　母线电压检测，完成电压闭环及保护作用。

e. 电流采样模块　输出电流检测，完成电流闭环及保护作用。

f. 并网电压采样模块　并网电压检测，完成电网电压锁相、电压前馈及保护作用。

g. 隔离驱动信号 1　完成对 Boost 电路模块开关管的隔离驱动作用。

h. 隔离驱动信号 2　完成对驱动电路 IPM 智能模块的隔离驱动作用。

i. 温度告警模块　IPM 智能模块的温度检测，完成并网逆变器过温保护功能。

j. DSP 控制电路　执行并网逆变器的软件算法功能。

k. 键盘接口模块　设置影响并网电流质量的参数。

l. 液晶显示模块　显示并网参数。

② 驱动电路模块　驱动电路模块将蓄电池电压通过 Boost 升压后，逆变成与电网电压同频、同相、同幅的正弦交流电，以实现与电网的并网连接。主要是三菱 IPM 智能模块的应用。

图 3.7.2 为驱动电路模块原理图，图 3.7.3 为驱动电路模块实物图，表 3.7.1 为驱动电路模块端口定义。

③ Boost 电路模块　Boost 电路模块主要将蓄电池组输出的 24V 直流电压变换成能满足并网要求的直流母线电压。

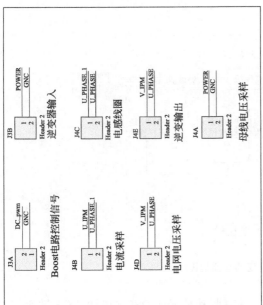
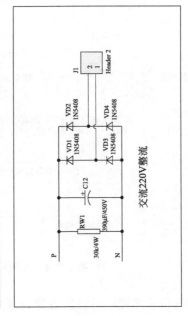
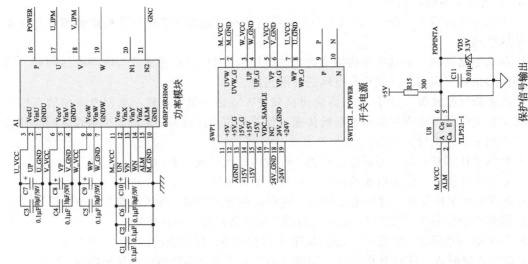
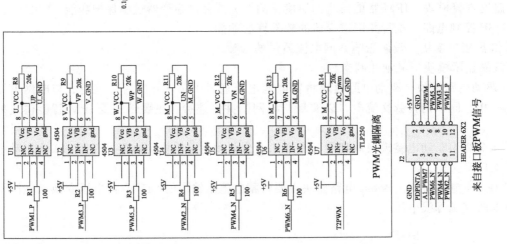

图 3.7.2 驱动电路模块原理图

第三部分 实训项目

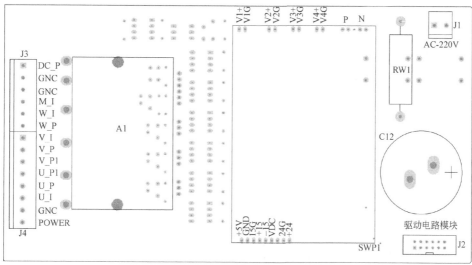

图 3.7.3 驱动电路模块实物图

表 3.7.1 驱动电路模块端口定义

序号	名称	说　明	扩展接口	备注
1	J1	交流 220V 电源		
2	J1			
3	J2	连到接口板的 J1 上		
4	J3：DC_P	电压升压控制信号输出		
5	J3：GNC			
6	J3：GNC	直流电压采样输出		
7	J3：M_I			
8	J3：W_I	逆变器电压输出		
9	J3：W_P			
10	J4：V_I	电网电压采样输出		
11	J4：V_P			
12	J4：V_P1	接电感挡板		
13	J4：U_P1			
14	J4：U_P	电流采样输出		
15	J4：U_I			
16	J4：GNC	逆变器电压输入		
17	J4：POWER			

图 3.7.4 为 Boost 电路模块原理图，图 3.7.5 为 Boost 电路模块实物图，表 3.7.2 为 Boost 电路模块端口定义。

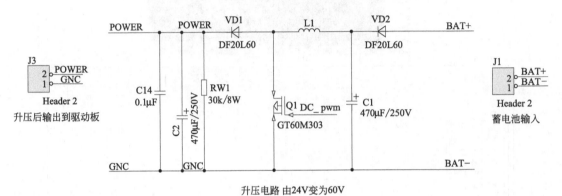

图 3.7.4　Boost 电路模块原理图

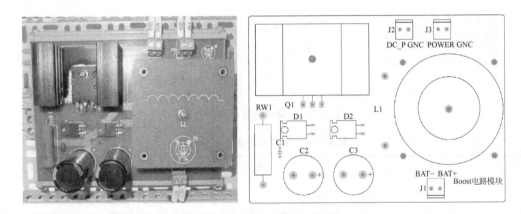

图 3.7.5　Boost 电路模块实物图

表 3.7.2　Boost 电路模块端口定义

序号	名称	说明	扩展接口	备注
1	J1:BAT−	蓄电池 24V 电源输入		
2	J1:BAT+			
3	J2:DC_P	电压升压控制信号输入		
4	J2:GNC			
5	J3:POWER	电压升压 60V 输出		
6	J3:GNC			

④ CPU 模块　CPU 模块完成并网逆变器的软件算法。CPU 模块采用 TI 公司的 32 位高性能定点 DSP 芯片 TMS320LF2812 作为核心芯片。

图 3.7.6 为 CPU 模块原理图，图 3.7.7 为 CPU 模块实物图，表 3.7.3 为 CPU 模块端口定义。

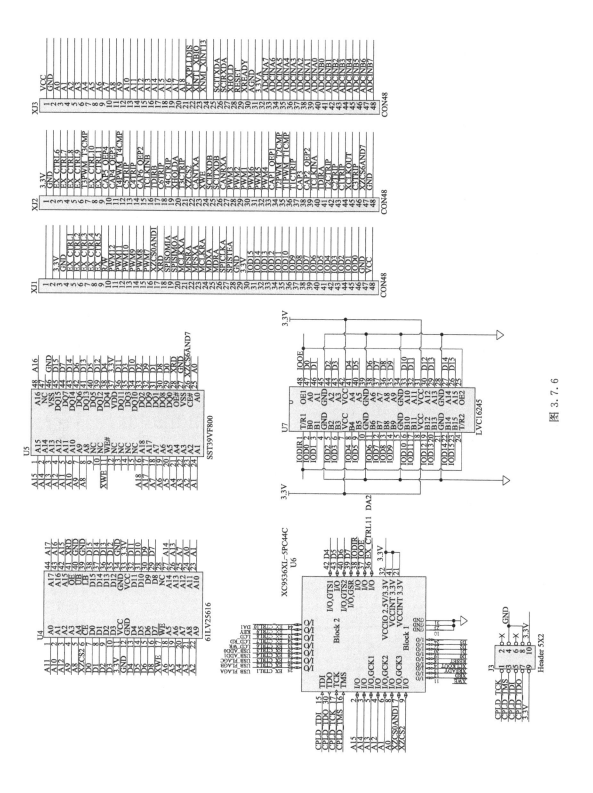

图 3.7.6

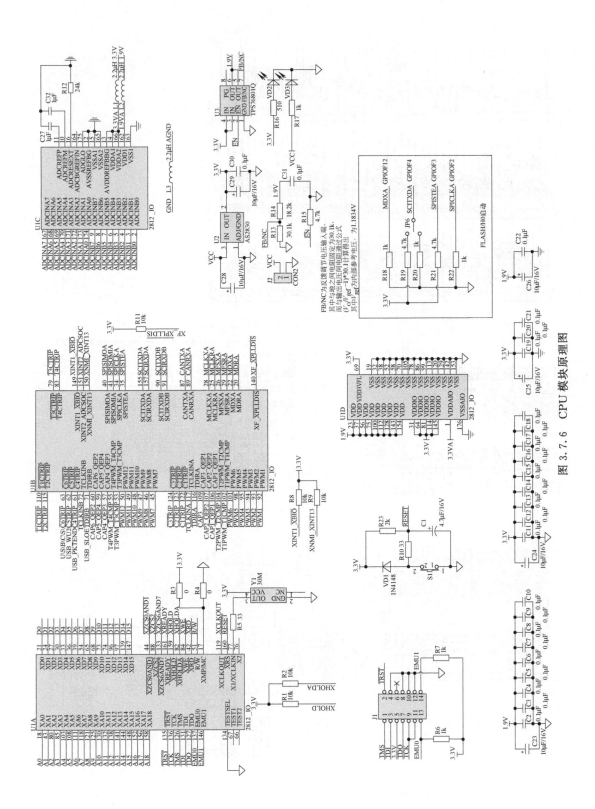

图 3.7.6 CPU 模块原理图

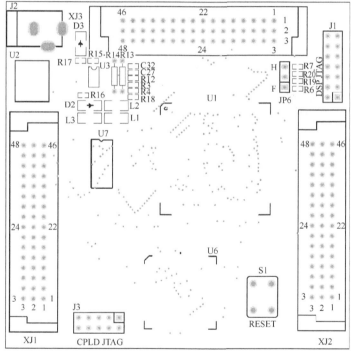

图 3.7.7 CPU 模块实物图

表 3.7.3 CPU 模块端口定义

序号	名称	说　　明	扩展接口	备注
1	J1	DSP JTAG		
2	J2	5V电源		调试用
3	J3	CPLD JTAG		
4	XJ1	插在接口板上		
5	XJ2			
6	XJ3			

⑤ 液晶显示模块 图 3.7.8 为液晶显示模块原理图，图 3.7.9 为液晶显示模块实物图，表 3.7.4 为液晶显示模块端口定义。

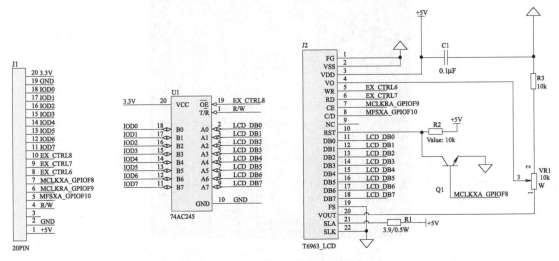

图 3.7.8　液晶显示模块原理图

图 3.7.9　液晶显示模块实物图

表 3.7.4　液晶显示模块端口定义

序号	名称	说明	序号	名称	说明
1	+5V	5V 电源	11	IOD7	数据位 7
2	GND	地	12	IOD6	数据位 6
3	空		13	IOD5	数据位 5
4	R/W	读/写信号	14	IOD4	数据位 4
5	GPIOF10		15	IOD3	数据位 3
6	GPIOF9		16	IOD2	数据位 2
7	GPIOF8		17	IOD1	数据位 1
8	EX_CTRL6		18	IOD0	数据位 0
9	EX_CTRL7		19	GND	地
10	EX_CTRL8		20	3.3V	3.3V 电源

⑥ 接口模块 图 3.7.10 为接口模块原理图，图 3.7.11 为接口模块实物图，表 3.7.5 为接口模块端口定义。

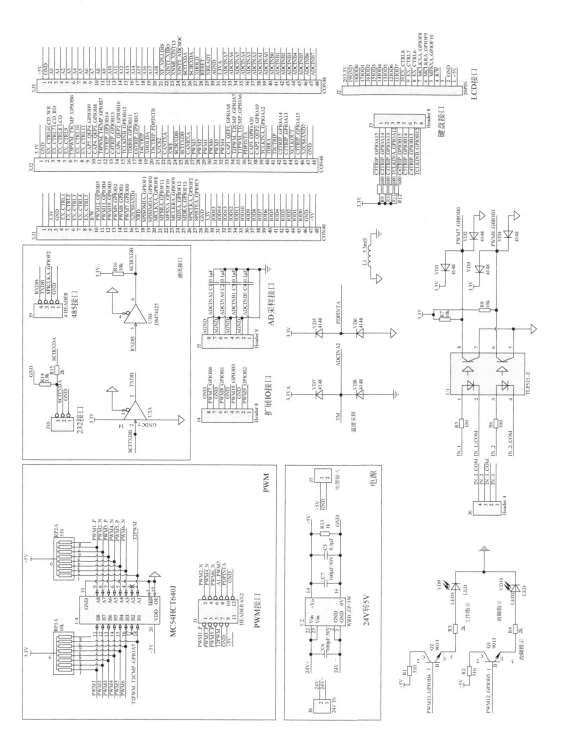

图 3.7.10 接口模块原理图

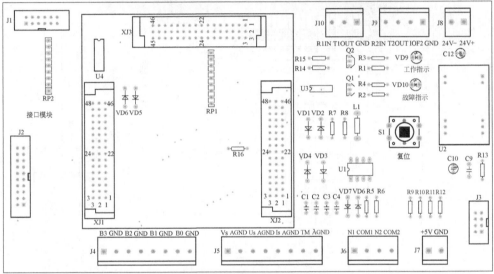

图 3.7.11 接口模块实物图

表 3.7.5 接口模块端口定义

序号	名称	说　明	扩展接口	备注
1	J1	至驱动模块	√	14P 排线
2	J2	至液晶显示模块	√	20P 排线
3	J3	至键盘接口模块	√	8P 排线
4	J4:B3	外扩 I/O 口	√	
5	J4:GND		√	
6	J4:B2	外扩 I/O 口	√	
7	J4:GND		√	
8	J4:B1	外扩 I/O 口	√	
9	J4:GND		√	
10	J4:B0	外扩 I/O 口	√	
11	J4:GND		√	
12	J5:Vs	直流电压采样输入,3.3V 以内电压信号		
13	J5:AGND			

续表

序号	名称	说　　明	扩展接口	备注
14	J5:Us	交流电压采样输入,3.3V 以内电压信号		
15	J5:AGND			
16	J5:Is	电流采样输入,3.3V 以内电压信号		
17	J5:AGND			
18	J5:TM	温度采样输入,3.3V 以内电压信号		
19	J5:AGND			
20	J6:N1	外扩 I/O 口	√	
21	J6:COM1		√	
22	J6:N2	外扩 I/O 口	√	
23	J6:COM2		√	
24	J7:+5V	隔离电源 DC/DC 输出 5V 电源正极		
25	J7:GND	隔离电源 DC/DC 输出 5V 电源负极		
26	J8:24V−	隔离电源 DC/DC 输入 24V 负极		
27	J8:24V+	隔离电源 DC/DC 输入 24V 正极		
28	J9:R2IN	485 输出信号	√	
29	J9:T2OUT		√	
30	J9:IOF2		√	
31	J9:GND		√	
32	J10:R1IN	232 输出信号	√	
33	J10:T1OUT		√	
34	J10:GND		√	
35	XJ1	插核心板		
36	XJ2			
37	XJ3			

⑦ 键盘接口模块　图 3.7.12 为键盘接口模块原理图,图 3.7.13 为键盘接口模块实物图,表 3.7.6 为键盘接口模块端口定义。

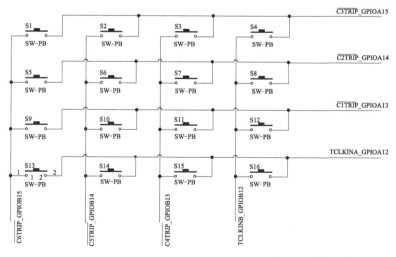
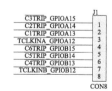

图 3.7.12　键盘接口模块原理图

 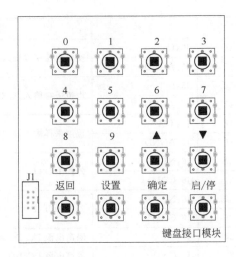

图 3.7.13 键盘接口模块实物图

表 3.7.6 键盘接口模块端口定义

序号	名称	说明	扩展接口	备注
1	J1	接到接口板 J3		

⑧ 母线电压采样模块　图 3.7.14 为母线电压采样模块原理图，图 3.7.15 为母线电压采样模块实物图，表 3.7.7 为母线电压采样模块端口定义。

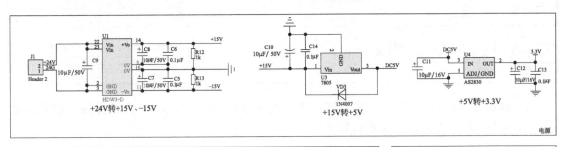

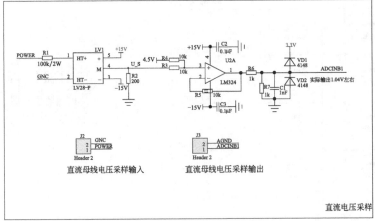 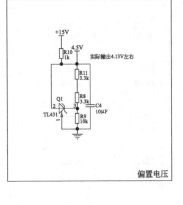

图 3.7.14 母线电压采样模块原理图

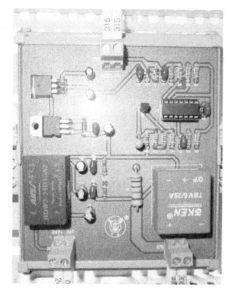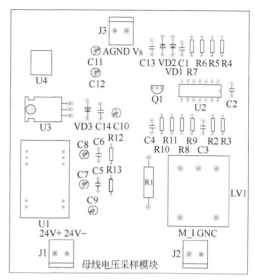

图 3.7.15　母线电压采样模块实物图

表 3.7.7　母线电压采样模块端口定义

序号	名称	说　　明	扩展接口	备注
1	J1:24V+	隔离电源 DC/DC 24V 输入		
2	J1:24V−			
3	J2:M_I	直流电压采样输入		
4	J2:GNC			
5	J3:Vs	电压采样调理输出		
6	J3:AGND			

⑨ 并网电压采样模块　图 3.7.16 为电网电压采样模块原理图，图 3.7.17 为电网电压采样模块实物图，表 3.7.8 为电网电压采样模块端口定义。

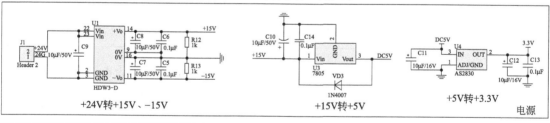

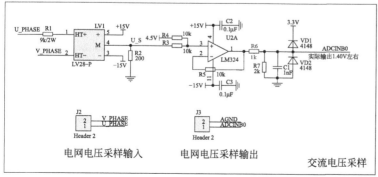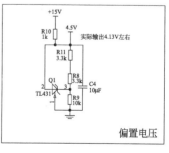

图 3.7.16　并网电压采样模块原理图

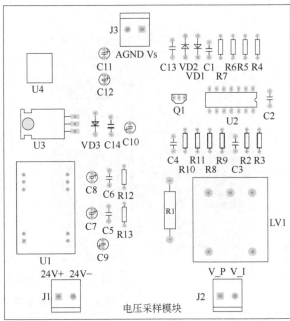

图 3.7.17　电网电压采样模块实物图

表 3.7.8　电网电压采样模块端口定义

序号	名称	说　明	扩展接口	备注
1	J1:24V+	隔离电源 DC/DC 24V 输入		
2	J1:24V−			
3	J2:V_P	交流电压采样输入		
4	J2:V_I			
5	J3:Vs	电压采样调理输出		
6	J3:AGND			

⑩ 电流采样模块　图 3.7.18 为电流采样模块原理图，图 3.7.19 为电流采样模块实物图，表 3.7.9 为电流采样模块端口定义。

表 3.7.9　电流采样模块端口定义

序号	名称	说　明	扩展接口	备注
1	J1:24V+	隔离电源 DC/DC 24V 输入		
2	J1:24V−			
3	J2:U_P	电流采样输入		
4	J2:U_I			
5	J3:Is	电流采样调理输出		
6	J3:AGND			

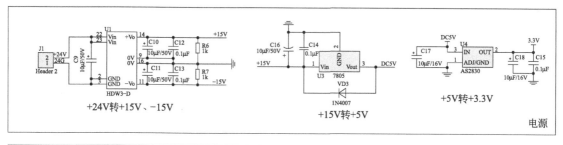
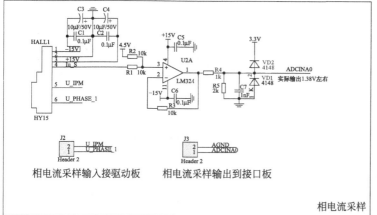
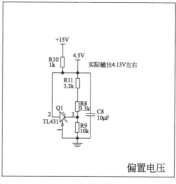

图 3.7.18　电流采样模块原理图

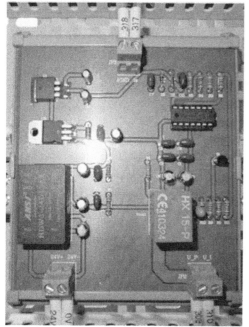
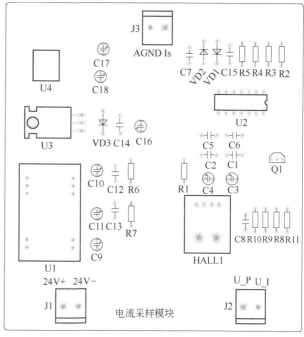

图 3.7.19　电流采样模块实物图

⑪ 滤波板　滤波板是一个由电感线圈构成的滤波器。图3.7.20为滤波板原理图，图3.7.21为滤波板实物图，表3.7.10为滤波板端口定义。

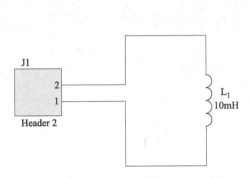

图3.7.20　滤波板原理图

图3.7.21　滤波板实物图

表3.7.10　滤波板端口定义

序号	名称	说　　明	扩展接口	备注
1	J1	滤波电感		
2	J1			

(2) 并网逆变器参数设置操作说明

并网逆变器

控制器上电初始，系统初始化，接口模块的工作指示灯和故障指示灯全亮，液晶屏显示"初始化…"，3~5s后，指示灯灭。液晶屏显示如图3.7.22所示界面。

① 参数设定步骤

a. 点击键盘上"确定"按键，出现参数设定界面，如图3.7.23所示。

图3.7.22　系统默认界面

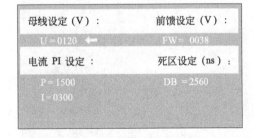

图3.7.23　参数设定界面

b. 通过键盘上的"▼"或"▲"按键，液晶屏上光标移动，选择要设置的参数，默认选择母线设定，点击"设置"键，出现如图3.7.24所示界面。

c. 通过键盘上的数字键设定需要的值，例如输入"1""0""0"，显示$U=100$，点击"确认"，完成参数设定，如图3.7.25所示。

d. 其他参数设定类似与母线设定，操作功能示意如图3.7.26所示。

② 参数设置说明　为了保证系统能够安全运行，参数设置有效范围做如下限定：

并网逆变器参数设置

a. 母线电压有效设置范围　60~120V；

图 3.7.24 参数设定界面（一）

图 3.7.25 参数设定界面（二）

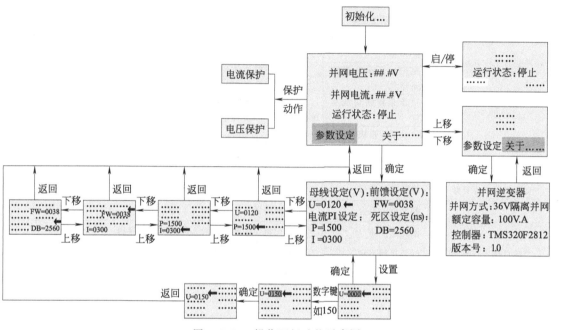

图 3.7.26 操作面板功能示意图

b. 电流环比例系数 P 有效设置范围　150～1500；

c. 电流环积分系数 I 有效设置范围　30～300；

d. 前馈电压有效设置范围　0～50V；

e. 死区设置有效值　2560ns、2780ns、2990ns、3200ns。

(3) 常见故障处理（表 3.7.11）

表 3.7.11　常见故障处理表

故障现象	故障的可能原因	故障的排除方法和步骤
供电异常	保险丝损坏	查看保险丝座内的保险丝是否烧毁。如烧毁，则更换新保险丝
	存在短路或者漏电	仔细检查线路是否存在短路,并用万用表测量各相之间阻值和各相间对地阻值是否正常。如果阻值为零或阻值很小,说明存在短路现象,应采取逐次断开的方法检查
逆变器上电异常	断路器损坏	检查直流负载断路器输入与输出是否导通
	线路接触不良	用万用表检查线路连接情况
	断路器未闭合	闭合断路器

续表

故障现象	故障的可能原因	故障的排除方法和步骤
闭合并网发电断路器后电压表没有显示	断路器损坏	检查直流负载断路器输入与输出是否导通
	线路接触不良	用万用表检查线路连接情况
	断路器未闭合	闭合断路器
	熔断芯未装	装入3A熔断芯
	熔断芯坏	更换熔断芯
逆变器工作异常	蓄电池断路器损坏	检查蓄电池断路器输入与输出是否导通
	蓄电池断路器未闭合	闭合蓄电池断路器
	没有启动逆变器	按键盘接口模块"启/停"键启动逆变器
	母线电压采样模块线路异常	用万用表检查母线电压采样模块线路连接情况,测试 U_s 和 AGND 之间的电压为 1.04V 左右
	电网电压采样模块线路异常	用万用表检查电网电压采样模块线路连接情况,测试 V_s 和 AGND 之间的电压为 1.38V 左右
	电流采样模块线路异常	用万用表检查电流采样模块线路连接情况,测试 I_s 和 AGND 之间的电压为 1.38V 左右
	液晶显示模块异常 键盘接口模块异常	检查连接排线是否松动、损坏
直流负载工作异常	断路器损坏	检查直流负载断路器输入与输出是否导通
	线路接触不良	用万用表检查线路连接情况
	断路器未闭合	闭合断路器
交流负载工作异常	断路器损坏	检查交流负载断路器输入与输出是否导通
	线路接触不良	用万用表检查线路连接情况
	断路器未闭合	闭合断路器

【项目原理及基础知识】

(1) 逆变器的控制方式

并网逆变器按控制方式分类,可分为电压源电压控制、电压源电流控制、电流源电压控制和电流源电流控制四种方式。但由于逆变回路中大电感往往会导致系统动态响应差,当前世界范围内大部分逆变器均采用以电压源输入为主。

逆变器与市电并联运行的输出控制可分为电压控制和电流控制。市电系统可视为容量无穷大的定值交流电压源,如果并网逆变器的输出采用电压控制,则实际上就是一个电压源与电压源并联运行的系统,这种情况下很难保证系统稳定运行;如果并网逆变器的输出采用电流控制,则只需控制并网逆变器的输出电流以跟踪市电电压,即可达到并联运行的目的。

并网逆变器一般都采用电压源输入、电流源输出的控制方式。

(2) 逆变器的工作原理

单相电压全控型 PWM 逆变器工作原理如图 3.7.27 所示,为通常使用的单相输出的全桥逆变主电路,其中交流元件采用 IGBT 管 Q11、Q12、Q13、Q14,并由 PWM 脉宽调制控制 IGBT 管的导通或截止。

当逆变器电路接上直流电源后,先由 Q11、Q14 导通,Q12、Q13 截止,则电流由直流电源正极输出,经 Q11、L、变压器初级线圈、Q14 回到电源负极。当 Q11、Q14 截止后,Q12、Q13 导通,电流从电源正极经 Q13、变压器初级线圈、L、Q12 回到电源负极。此时,

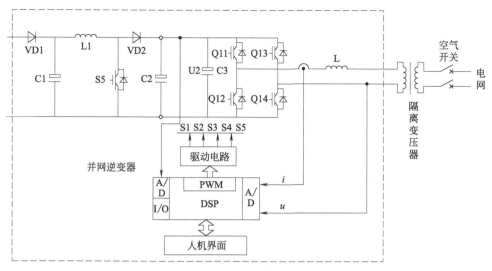

图 3.7.27 逆变器主电路

在变压器初级线圈上已形成正负交变的方波,利用高频 PWM 控制,两对 IGBT 管交替重复,在变压器上产生交流电压。由于 LC 交流滤波器的作用,使输出端形成正弦波交流电压。

随着新能源的发展应用,分布式发电系统的并网装置——并网逆变器的研究越来越受到关注,并网逆变器是完成发电系统产生的直流电能到适合于电网使用的交流电能的转换装置。然而随着投放使用的并网逆变装置增多,其输出的进网电流谐波对电网的污染也不容忽视,通常采用进网电流总的谐波畸变率(THD)来描述电能质量。进网电流中的谐波除了会给电网造成额外的损耗外,还可能损坏电网中的用电设备。本并网逆变器控制采用电压型 PWM 技术,完成对电网的锁相、直流母线调节和电流调节,实现电能并网和带负载运行,具有正弦波电流输出和单位功率因数的特点,有效地解决了并网装置的谐波污染、功率因数低等问题。

(3) 电流控制策略

采用电压外环、电流内环控制,首先建立两相同步旋转坐标系下逆变器数学模型,在此基础上给出了基于空间矢量调制的电流闭环控制策略,实现了并网电流有功分量和无功分量的独立控制。电流内环完成并网电流相位及幅值的控制,即跟踪并网电流指令 i_{ref}。电流内环一般采用加入网电压前馈的 PI 控制器,电流反馈采样值和电流指令值比较,其误差通过 PI 控制器,输出指令类型为电压,与电网前馈电压相加后,得到所需的并网电压指令。加入网电压前馈,实际上是抵消了电网电压,使得电流 PI 环输出值即电感电压,进而微调控制并网电流,是一种超前控制。在电压前馈 PI 控制中,由于 PI 控制器在跟踪正弦信号时不可避免会出现稳态误差,使得实际输出电流无法与输出电流指令相同。

电流内环控制一般有电流滞环跟踪控制和恒定开关频率电流控制两种,本装置采用后者,原理框图如图 3.7.28 所示,正弦电流基准值 i_{ref} 和输出瞬时电流 i_o 比较得到误差量,进入控制器调节后加前馈电压送到比较器,与三角波比较得到 SPWM 信号去控制主电路功率管的导通与截止。本装置产生 SPWM 的方式选择单极性倍频正弦脉宽调制(SPWM)方式,在不提高开关频率的前提下,提高了 SPWM 波形的谐波频率,从而使输出电压的谐波分量可以得到有效控制。

电压外环的作用主要是通过 PI 控制器实现对直流电压指令 U_{dc} 的跟踪,同时给出电流

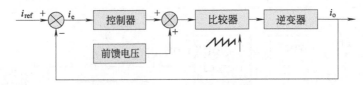

图 3.7.28　电流控制策略原理框图

内环指令 i_{ref}，其中直流电压指令 U_{dc} 是固定常数，通过前端 Boost 电路提供的母线电压 U 与其比较调节产生电流给定指令 i_{ref}。

(4) 影响电流并网质量的其他参数

并网逆变器的供电电能质量除了与控制策略有关外，还受并网开关闭合时产生的进网电流环流、逆变器开关死区时间、直流电压、电网扰动及隔离变压器铁芯的饱和非线性特性等影响。

当并网开关闭合之前，逆变器的输出电压与电网电压不一致，包括幅值、相位、频率、直流分量等不一致，即使此时进网电流给定值为 0，也会在逆变器与电网之间有电流流过，这里称其为环流。环流包括直流环流、基波环流和谐波环流，环流在轻载时占的比例较大，重载时占的比例减小。由于逆变器的等效输出阻抗和线路阻抗很小，所以逆变器输出电压与电网电压微小的电压差都会带来很大的环流，并且在逆变器与电网之间有电感存在，使环流滞后电压 90°，所以该环流的存在严重影响了网侧电流质量。逆变器的控制采用 PI 调节，而 PI 调节并不能实现无静差的跟踪，所以逆变器输出电压与电网电压之间存在幅值差和相位差，减小环流的问题就转化为减小 PI 调节稳态误差的问题。

当逆变器控制开关频率较高时，死区时间对输出电流波形影响不可忽略。

电网电压扰动是一个暂态过程，其影响不是持续的，而变压器受到制造及运行条件等因素影响，实际运行时铁芯非线性饱和常常会产生较强的谐波电流扰动，且扰动是持续的，对输出电流波形质量产生较大的影响。

本实训采用数字锁相方法，正弦表格由 4000 个点构成，则锁相精度为 360/4000＝0.09，cos0.09＝0.99999，所以该锁相方法能获得很好的锁相效果，从而有较高的 PF 值。锁相采用在旋转坐标下跟踪电网电压合成矢量来实现，亦即求出电网电压合成矢量的空间位置角，作为控制策略中所用坐标变换的给定角度，便可实现电网相位与相序的跟踪。

并网型逆变器工作操作演示及电能质量分析

并网逆变器工作实训

【项目实施】

① 打开各控制系统总电源开关，电源指示灯有显示。

② 将充放电控制器的刹车置于"RELEASE（自动刹车）"状态，合上"能源转换储存控制系统"的"蓄电池"空气开关，蓄电池接入光伏 MPPT 控制系统，同时给充放电控制器供电，此时充放电控制器进行初始化，红色指示灯点亮（工作在刹车状态）。必须等红色指示灯熄灭（退出刹车状态）才能进行下一步操作。

③ 打开"模拟能源控制系统"的"变频器"开关，给变频器上电。

④ 再按下控制开关部分的"启动"按钮（绿色按钮），操作变频器选择合适的频率，按 RUN 键运行模拟风能装置（频率不大于 20Hz）。

⑤ 打开"并网逆变控制系统"的"控制器"开关，并网逆变控制器上电，液晶屏初始化。

⑥ 依次打开"蓄电池"和"并网发电"开关，逆变器输入和输出电压部分有数值。

⑦ 操作人机界面，通过键盘移动光标来选择"参数设定"母线电压

$U=120\text{V}$,电流环比例系数 $P=1500$,电流环积分系数 $I=300$,前馈电压 $FW=38\text{V}$,死区时间 $DB=2560\mu\text{s}$。点击"返回"键,返回初始界面,再点击"启/停"键,启动逆变器。记录逆变器工作并网前后各电表的值于下表。

序号	项目	逆变器输入		逆变器输出	
		U/V	I/A	U/V	I/A
1	并网前				
2	并网后				

⑧ 逆变器正常工作后,分别记录空载、直流负载、交流负载时逆变器输出电量表的参数于下表。

序号	项目	逆变器输出电量表				
		U/V	I/A	P/kW	Q/kV·A	PF
1	并网后(空载)					
2	并网后(电机)					
3	并网后(LED灯)					

⑨ 操作人机界面,通过键盘移动光标来选择"参数设定"母线电压 $U=120\text{V}$,电流环比例系数 $P=150$,电流环积分系数 $I=30$,前馈电压 $FW=38\text{V}$,死区时间 $DB=2560\mu\text{s}$。点击"返回"键,返回初始界面,再点击"启/停"键,启动逆变器。通过改变电流环 PID 参数的设定,记录逆变器输出电量表的数值于下表,并画出相应的谐波波形。

序号	项目	逆变器输出电量表				
		U/V	I/A	PF	电压 THD	电流 THD
1	$P=150;I=30$					
2	$P=500;I=100$					
3	$P=800;I=180$					
4	$P=1200;I=250$					
5	$P=1500;I=300$					

⑩ 操作人机界面,通过键盘移动光标来选择"参数设定"母线电压 $U=120\text{V}$,电流环比例系数 $P=1500$,电流环积分系数 $I=300$,前馈电压 $FW=38\text{V}$,死区时间 $DB=2560\mu\text{s}$。点击"返回"键,返回初始界面。通过改变死区时间 DB 参数的设定,记录逆变器输出电量表的数值于下表。

序号	项目	逆变器输出电量表				
		U/V	I/A	PF	电压 THD	电流 THD
1	$DB=2560\mu\text{s}$					
2	$DB=2780\mu\text{s}$					
3	$DB=2990\mu\text{s}$					
4	$DB=3200\mu\text{s}$					

⑪ 实验结束后,点击"启/停"键使逆变器停止工作,然后操作变频器的"STOP"键,停止风机运行。

⑫ 依次操作"模拟能源控制系统"控制开关部分的"停止"按钮（红色按钮）和"变频器"空气开关。

⑬ 依次关闭"并网逆变控制系统"的"并网发电""直流负载""交流负载""蓄电池""控制器"空气开关和"能源转换储存控制系统"的"蓄电池"空气开关，最后再关闭控制系统的总电源开关。如需进行后续实验，可不关闭控制系统总电源。

【项目作业】

① 逆变器正常工作后，改变母线电压 U 的值来观测各电表值。

② 逆变器正常工作后，通过改变母线电压值观测电流谐波畸变率（THD）的值。

项目八　运行与调试光伏发电系统

【项目描述】

学习太阳能光伏发电技术的关键内容，了解太阳能电池发电的基本原理，通过实训平台，完成光伏发电系统的运行与调试。

【能力目标】

① 掌握并网光伏发电系统的原理。

② 掌握并网光伏发电系统中太阳能电池板、光伏控制器、蓄电池组、并网逆变器的作用。

【项目环境】

光伏发电是利用半导体界面的光生伏特效应而将光能直接转变为电能的一种技术。该系统主要由太阳电池板（组件）、控制器和逆变器三大部分组成，主要部件由电子元器件构成。太阳能电池经过串联后进行封装保护，可形成大面积的太阳电池组件，再配合功率控制器等部件，就形成了光伏发电装置。

光伏发电系统运行与调试

(1) 直流单元模块

见项目二【项目环境】(3)（见第 20 页）。

(2) 密封铅酸蓄电池组

技术指标

单节蓄电池容量：12V，24Ah

单节蓄电池尺寸：165mm×125mm×175mm

实物图（图 3.8.1）

端口定义（表 3.8.1）

表 3.8.1　蓄电池组端口定义

序号	定义	说明
1	BATX+	蓄电池正极输出
2	BATX−	蓄电池负极输出

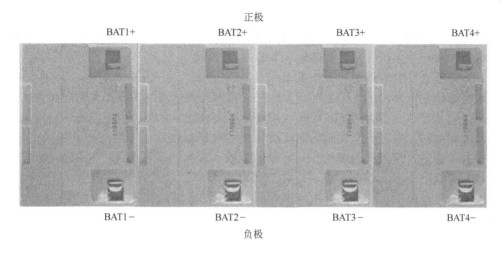

图 3.8.1 蓄电池组实物图

(3) MPPT 控制器

见项目五【项目环境】(1)（见第 41 页）。

【项目原理及基础知识】

(1) 太阳能光伏发电系统的构成

通过太阳能电池将太阳辐射能转换为电能的发电系统，称为太阳能光伏发电系统，也可叫太阳能电池发电系统。尽管太阳能光伏发电系统应用形式多种多样，应用规模也跨度很大，从小到不足 1W 的太阳能草坪灯，大到几百千瓦甚至几兆瓦的大型光伏发电站，但太阳能光伏发电系统的组成结构和工作原理却基本相同。其主要结构由太阳能电池组件（或方阵）、蓄电池（组）、光伏控制器、逆变器（在有需要输出交流电的情况下使用）以及一些测试、监控、防护等附属设施构成。

① 太阳能电池组件 太阳能电池组件也叫太阳能电池板，是太阳能发电系统中的核心部分，也是太阳能发电系统中价值最高的部分。其作用是将太阳光的辐射能量转换为电能，并送往蓄电池中存储起来，也可以直接用于推动负载工作。当发电容量较大时，就需要用多块电池组件串、并联后构成太阳能电池方阵。目前应用的太阳能电池主要是晶体硅电池，分为单晶硅太阳能电池、多晶硅太阳能电池和非晶硅太阳能电池等几种。

② 蓄电池 蓄电池的作用主要是存储太阳能电池发出的电能，并可随时向负载供电。太阳能光伏发电系统对蓄电池的基本要求是：自放电率低、使用寿命长、充电效率高、深放电能力强、工作温度范围宽、少维护或免维护以及价格低廉。目前为光伏系统配套使用的主要是免维护铅酸电池，在小型、微型系统中，也可用镍氢电池、镍镉电池、锂电池或超级电容器。当需要大容量电能存储时，就需要将多只蓄电池串、并联起来构成蓄电池组。

③ 光伏控制器 太阳能光伏控制器的作用是控制整个系统的工作状态，其功能主要有防止蓄电池过充电保护、防止蓄电池过放电保护、系统短路保护、系统极性反接保护、夜间防反充保护等。在温差较大的地方，控制器还具有温度补偿的功能。另外，控制器还有光控开关、时控开关等工作模式，以及充电状态、蓄电池电量等各种工作状态的显示功能。光伏控制器一般分为小功率、中功率、大功率和风光互补控制器等。

④ 交流逆变器 交流逆变器是把太阳能电池组件或者蓄电池输出的直流电转换成交流电，供应给电网或者交流负载使用的设备。逆变器按运行方式可分为独立运行逆变器和并网逆变器。独立运行逆变器用于独立运行的太阳能发电系统，为独立负载供电。并网逆变器用

于并网运行的太阳能发电系统。

⑤ 光伏发电系统附属设施　光伏发电系统的附属设施包括直流配线系统、交流配电系统、运行监控和检测系统、防雷和接地系统等。

(2) 太阳能光伏发电系统的工作原理

太阳能光伏发电系统从大类上可分为独立（离网）光伏发电系统和并网光伏发电系统两大类。图3.8.2是独立型太阳能光伏发电系统的工作原理示意图。太阳能光伏发电的核心部件是太阳能电池板，它将太阳光的光能直接转换成电能，存储于蓄电池中。当负载用电时，蓄电池中的电能通过控制器合理地分配到各个负载上。太阳能电池所产生的电流为直流电，可以直接以直流电的形式应用，也可以用交流逆变器将其转换成为交流电，供交流负载使用。太阳能发电的电能可以即发即用，也可以用蓄电池等储能装置将电能存储起来，在需要时使用。

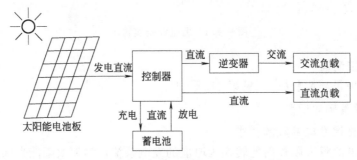

图3.8.2　独立型太阳能光伏发电系统工作原理

图3.8.3是并网型太阳能光伏发电系统工作原理示意图。并网型光伏发电系统由太阳能电池组件方阵将光能转变成电能，并经直流配线箱进入并网逆变器，有些类型的并网型光伏系统还要配置蓄电池组存储直流电能。并网逆变器由充放电控制、功率调节、交流逆变、并网保护切换等部分构成。经逆变器输出的交流电供负载使用，多余的电能通过电力变压器等设备馈入公共电网（可称为卖电）。当并网光伏系统因天气原因发电不足或自身用电量偏大时，可由公共电网向交流负载供电（称为买电）。系统还配备有监控、测试及显示系统，用于对整个系统工作状态的监控、检测及发电量等各种数据的统计，还可以利用计算机网络系统远程传输控制和显示数据。

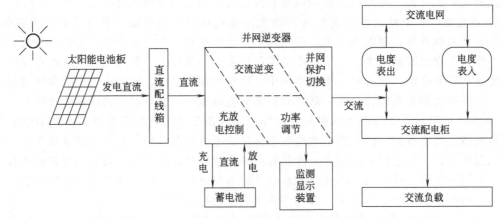

图3.8.3　并网型太阳能光伏发电系统工作原理

【项目实施】

① 合上各控制柜上的总电源开关，系统得电，三相电源指示灯亮。

② 合上"模拟能源控制系统"的"开关电源"空气开关，使开关电源工作。

③ 合上"模拟光源"空气开关，使模拟太阳光灯打开。

④ 合上"PLC"空气开关，PLC上电，将PLC上的拨动开关置"RUN"状态，使PLC程序处于运行状态。

光伏发电系统运行与调试操作演示

⑤ 按下追日系统控制开关的"开始"按钮（绿色按钮），各步进电机准备工作；按下"控制"按钮（黄色按钮），模拟光源开始运动。

⑥ 将充放电控制器的刹车置于"RELEASE（自动刹车）"状态，合上"能源转换储存控制系统"的"蓄电池"空气开关，蓄电池接入光伏MPPT控制系统，同时给充放电控制器供电，此时充放电控制器进行初始化，红色指示灯点亮（工作在刹车状态）。必须等红色指示灯熄灭（退出刹车状态）才能进行下一步操作。

⑦ 合上"光伏输出"和"光伏MPPT"空气开关，按"CPU核心模块"上的复位按钮K1，系统复位。

⑧ 按"人机交互模块"的"DOWN"键切换到"功率跟踪CVT"，再点击"ENTER"键确定，系统将自动进行功率跟踪。

⑨ 合上"并网逆变控制系统"的"控制器"空气开关，并网逆变器上电，液晶屏初始化。

⑩ 合上"蓄电池"和"并网发电"空气开关，各智能表有数值显示。

⑪ 按下"键盘接口模块"的"启/停"键，并网逆变器开始工作。工作指示灯亮。此时可选择打开"直流负载"和"交流负载"空气开关，观察逆变器输出功率的变化。

⑫ 记录"THNRFG-4型风光互补发电系统监控软件"中各智能表的数值于表3.8.2和表3.8.3。

表3.8.2 能源转换储存控制系统各电量表数据

光伏组件电量			蓄电池组电量		
U/V	I/A	P/W	U/V	I/A	P/W

表3.8.3 并网逆变控制系统各电量表数据

逆变器输入电量（直流）			逆变器输出电量（交流）						
U/V	I/A	P/W	U/V	I/A	P/W	Q/V·A	PF	电压THD	电流THD

⑬ 实验结束后，点击"并网逆变控制系统""键盘接口模块"的"启/停"键，使逆变器停止工作，然后点击"模拟能源控制系统"追日系统控制开关的"停止"按钮（红色按钮），停止追日系统。

⑭ 依次关闭"模拟能源控制系统"的"模拟光源""开关电源""PLC"空气开关。

⑮ 依次关闭"能源转换储存控制系统"的"光伏输出""光伏MPPT""蓄电池"的空气开关。

⑯ 依次关闭"并网逆变控制系统"的"并网发电""直流负载""交流负载""蓄电池"

"控制器"空气开关,最后再关闭各控制系统的总电源开关。如需进行后续实验,可不关闭控制系统总电源。

【项目作业】

画出实验平台的太阳能光伏发电系统工作原理框图,并写出能量传递过程。

项目九 运行与调试风光互补发电系统

【项目描述】

学习风光互补控制的关键技术,可以独立对系统并网逆变器参数进行设置,能够对电能质量进行分析。

【能力目标】

① 掌握风光互补发电系统的原理。
② 掌握风光互补发电系统各组成部分的作用。

【项目环境】

风光互补发电系统运行与调试由模拟能源控制系统、能源转换储存控制系统、并网逆变控制系统和能源监控管理系统组成。

① 模拟能源控制系统由控制屏(电源、网孔板、工具抽屉组成)、可编程序控制器(PLC)、编程线、模拟量模块、变频器、触摸屏、交流接触器、继电器、按钮、开关等组成,参阅表3.9.1。

风光互补发电系统运行与调试

表3.9.1 模拟能源控制系统

序号	名　称	型号、规格说明	数量	备注
1	控制屏	800mm×600mm×1880mm	1套	
2	可编程序控制器(PLC)	FX3U-32MT 晶体管主机 输入:16点;输出:16点	1台	三菱
3	模拟量模块	FX3U-3A-ADP 输入通道:2个;输出通道:1个 最大分辨率:12位 范围:DC0～5V/10V,DC4～20mA	1块	三菱
4	变频器	FR-A800 额定功率:0.75kW 额定输入电压:3相,AC380～480V,50/60Hz 额定输出电压:3相,AC380～480V,输出频率:0.2～400Hz	1台	三菱

② 能源转换储存控制系统由控制屏(电源、网孔板、工具抽屉组成)、光伏阵列汇流模块、直流电源防雷器、直流电压智能数显表、直流电流智能数显表、磁盘电阻器、直流电压电流采集模块、CPU核心模块、人机交互模块、PWM驱动模块、通信模块、无线通信模块、温度告警模块、DC-DC Boost/Buck/Boost-Buck三种主电路模块、蓄电池组、充放电控制器、51 ISP下载器、PIC编程器等组成,参阅表3.9.2。

表 3.9.2 能源转换储存控制系统

序号	名称	型号、规格说明	数量	备注
1	控制屏	800mm×600mm×1880mm	1套	
2	直流电压智能数显表	输入范围:0~600V;精度:0.5级	2只	RS-485接口
3	直流电流智能数显表	输入范围:0~5A;精度:0.5级	2只	RS-485接口
4	光伏阵列汇流模块	4路	1块	
5	直流电源防雷器	工作电压:24V DC;放电电流:5kA	1只	
6	磁盘电阻器	2×72Ω,1.45A	1个	
7	CPU核心模块	CPU芯片:AT89S52 支持ISP下载 模拟量输入通道:8路 开关量输入输出:8路 串行通信接口:1路 人机交互接口:1路	1块	
8	人机交互模块	独立按键:5个; 12864字符型液晶	1块	
9	PWM驱动模块	CPU芯片:PIC16F690 串行通信口:1路 开关量输入输出:8路 隔离PWM驱动信号:2路	1块	
10	通信模块	RS-232接口:2路 RS-485接口:2路	1块	
11	无线通信模块	ZigBee频段:2405~2480MHz;传输距离:>10m Bluetooth频段:2.402~2.480GHz;可同时支持多个蓝牙设备 WiFi支持802.11b/g/n无线标准 支持透明/协议两种数据传输模式;支持心跳信号;WiFi连接指示	2块	
12	温度告警模块	光耦隔离继电器输出:3路 温度传感器:1路	1块	
13	直流电压电流采集模块	电压通道:1路 电流通道:1路	2块	
14	DC-DC主电路模块	Boost主电路	1块	
		Buck主电路	1块	
		Boost-Buck主电路	1块	
15	蓄电池组	12V/24Ah密封铅酸蓄电池	4块	
16	充放电控制器	额定电压:12V、24V自动识别 最大负载电流:35.61A/17.80A 蓄电池充满断开电压:0~15V/30V可设置	1只	
17	51 ISP下载器	支持89S51/52在线下载	1套	
18	PIC编程器	支持PIC系列芯片烧录	1套	

③ 并网逆变控制系统由DSP核心模块、接口模块、液晶显示模块、键盘接口模块、驱动电路模块、Boost电路模块、母线电压采样模块、电网电压采样模块、电流采样模块、直流负载、交流负载、直流电压智能数显表、直流电流智能数显表、逆变输出电量表、隔离变压器、离网逆变器、DSP仿真器等组成,参阅表3.9.3。

并网逆变将DC24V逆变成AC36V、50Hz,经变压器升至AC220V与单相市电并网。主控制器采用TI定点32位TMS320F2812芯片,输出功率因数接近于1。采用双闭环控制,内环为电流环,外环为电压环,并网同步采用数字锁相技术。

表 3.9.3 并网逆变控制系统

序号	名 称	型号、规格说明	数量	备注
1	控制屏	800mm×600mm×1880mm	1套	
2	直流电压智能数显表	输入范围：0~600V；精度：0.5级	1只	RS-485接口
3	直流电流智能数显表	输入范围：0~5A；精度：0.5级	1只	RS-485接口
4	逆变输出电量表	输入网络：单相2线 输入频率：45~65Hz 输入电压额定值：AC100V 输入电流额定值：AC5A 测量精度：频率0.05Hz，无功电度表1级，其他0.5级	1只	RS-485接口
5	直流负载	额定电压：DC24V	1只	
6	交流负载	额定电压 AC/DC 36V	1只	
7	隔离变压器	输入电压 36V，输出电压 220V 频率 50Hz	1只	
8	驱动电路模块	IPM智能功率模块：6MBP20RH060 额定电流：20A 直流母线电压：450V 最大开关频率：20kHz	1块	
9	Boost电路模块	直流升压	1块	
10	DSP核心模块	CPU：TMS320LF2812 主频：150MHz RAM：4Mb(256K×16Bit) Flash：8Mb(512K×16Bit)	1块	
11	液晶显示模块	240×128 液晶屏	1块	
12	接口模块	模拟量输入通道：4路 串行通信接口：1路 键盘接口：1路 液晶屏接口：1路	1块	
13	键盘接口模块	4×4 矩阵式键盘	1块	
14	母线电压采样模块	电压采集通道：1路	1块	
15	电网电压采样模块	电压采集通道：1路	1块	
16	电流采样模块	电流采集通道：1路	1块	
17	离网逆变器	输入电压：DC24V 效率：>85% 输出电压：AC220V 额定功率：300W 输出波形：正弦波 输出频率：50Hz 波形失真：3%	1只	
18	DSP仿真器	支持DSP在线仿真、下载	1套	

④ 能源监控管理系统由系统控制器核心模块、继电器模块、15in工业平板电脑、组态软件等组成。能源监控管理系统可与各控制系统通信，上位机软件可实时显示运行数据，并可根据控制要求自动或手动改变运行状态。参阅表3.9.4。

表 3.9.4 能源监控管理系统

序号	名 称	型号、规格说明	数量	备注
1	控制屏	800mm×600mm×1880mm	1套	
2	15in工业平板电脑	显示器：15in液晶 CPU：凌动 N270,1.6GHz 主板：微星 A9830IMS 内存：DDR2 1G 硬盘：160G 触摸屏：4线电阻触摸屏 网卡：10~100M 串口：5个 USB：4个 VGA接口：1个 PCI插槽：1个	1台	

续表

序号	名 称	型号、规格说明	数量	备注
3	系统控制器核心模块	模拟量输入通道:3路 模拟量分辨率:8位 开关量输出:5路 支持 ISP 下载 串行通信接口:1路 蓄电池电量监测指示 支持蓄电池电量模拟	1块	
4	继电器模块	光耦隔离继电器输出:5路	1块	
5	组态软件	工业组态软件	1套	

【项目原理及基础知识】

风力发电和太阳能发电一样都受天气环境影响较大，带有一定的局限性，但它们之间存在一定的互补性。所谓"风光互补"是指在白天、夜间交替使用太阳能发电和风力发电。一般来说白天晴天，风可能比较小，以太阳能发电为主，以风力发电为辅；而夜间只能靠风机发电，往往风力也比白天大，从而形成全天的互补发电形式。使用适当比例的太阳能和风力发电，将获得最佳的投资组合。具体组合比例要经过技术、投资、气象、负荷情况综合论证。"风光互补"发电虽然能构成一定的互补关系，但仍受气象条件影响较大，如果加装蓄电池则能显著改善其稳定性。但蓄电池属于储能元件，使用它储能虽可在一定程度上弥补两者供电的不稳定性，可储存的能量毕竟有限，不能长时间持续供给功率，如遇连续阴雨天气和连续无风天气，整个供电系统的供电能力将会大大下降。对于比较重要的或供电稳定性要求较高的负载，还需考虑采用备用的柴油发电机组，形成风机、光伏和柴油发电机一体化的供电系统，供电的可靠性和稳定性将大为提高。

【项目实施】

① 合上各控制柜上的总电源开关，系统得电，三相电源指示灯亮。

② 合上"模拟能源控制系统"的"开关电源""模拟光源""变频器"空气开关。

③ 合上"PLC"空气开关，PLC 上电，将 PLC 上的拨动开关置"RUN"状态，使 PLC 程序处于运行状态。

风光互补发电系统运行与调试操作演示

④ 按下追日系统控制开关的"开始"按钮（绿色按钮），各步进电机准备工作；按下"控制"按钮（黄色按钮），模拟光源开始运动。

⑤ 将充放电控制器的刹车置于"RELEASE（自动刹车）"状态，合上"能源转换储存控制系统"的"蓄电池"空气开关，蓄电池接入光伏 MPPT 控制系统，同时给充放电控制器供电，此时充放电控制器进行初始化，红色指示灯点亮（工作在刹车状态）。必须等红色指示灯熄灭（退出刹车状态），才能进行下一步操作。

⑥ 合上"光伏输出"和"光伏 MPPT"空气开关，按"CPU 核心模块"的复位按钮K1，系统复位。

⑦ 按"人机交互模块"的"DOWN"键切换到"功率跟踪 CVT"，再点击"ENTER"键确定，系统将自动进行功率跟踪。

⑧ 按下"模拟能源控制系统"的绿色启动按钮，给变频器上电（当充放电控制器的红色指示灯亮时，无法启动变频器，因为充放电控制器工作在刹车状态）。按 PU/EXT 键，进入 PU 运行模式，通过 M 旋钮将频率设定在 20Hz 以下，按 RUN 键运行。

⑨ 合上"并网逆变控制系统"的"控制器"空气开关，并网逆变器上电，液晶屏初始化。

⑩ 合上"蓄电池"和"并网发电"空气开关，各智能表有数值显示。

⑪ 记录"THNRFG-4 型风光互补发电系统监控软件"中各智能表的数值于表 3.9.5 和表 3.9.6。

表 3.9.5 能源转换储存控制系统各电量表数据

光伏组件电量			蓄电池组电量		
U/V	I/A	P/W	U/V	I/A	P/W

表 3.9.6 并网逆变控制系统各电量表数据

逆变器输入电量(直流)			逆变器输出电量(交流)						
U/V	I/A	P/W	U/V	I/A	P/W	Q/V·A	PF	电压 THD	电流 THD

⑫ 实验结束后,点击"并网逆变控制系统""键盘接口模块"的"启/停"键,使逆变器停止工作;然后操作变频器的"STOP"键,停止风机运行;最后点击"模拟能源控制系统""追日系统控制开关"的"停止"按钮(红色按钮),停止追日系统。

⑬ 操作"模拟能源控制系统"控制开关部分的停止按钮(红色按钮),断开变频器电源;依次关闭"模拟能源控制系统"的"变频器""模拟光源""开关电源""PLC"空气开关。

⑭ 依次关闭"能源转换储存控制系统"的"光伏输出""光伏 MPPT""蓄电池"空气开关。

⑮ 依次关闭"并网逆变控制系统"的"并网发电""蓄电池""控制器"空气开关,最后再关闭各控制系统的总电源开关。如需进行后续实验,可不关闭控制系统总电源。

【项目作业】

① 画出实验平台的风光互补发电系统工作原理框图,并写出能量传递过程。
② 说明光伏发电与风光互补发电系统的组成和优缺点。

项目十 能源监控管理系统组态设计

【项目描述】

主要完成实时数据库组态、采集系统组态、数据服务组态、画面组态、网络站点组态、报表开发、脚本编程以及常用的功能参数配置工作。

【能力目标】

① 掌握旋思组态软件的使用方法。
② 能够独立完成软件的二次开发。

【项目环境】

(1) 能源监控管理单元

能源监控管理单元主要是针对电池电量、风光互补系统工作状态等整个过程的监测与控制。

能源监控管理

(2) 能源监控管理单元的组成

能源监控管理单元功能框图如图 3.10.1 所示,由系统控制器模块、继电器模块、通信

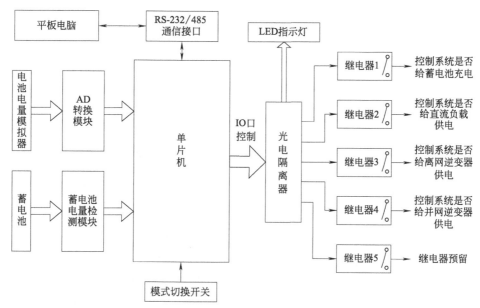

图 3.10.1　能源监控管理单元功能框图

接口模块和风光互补发电系统监控软件组成。系统控制器模块部分能够实时显示电池电量，进行风光互补系统工作状态切换，还可将输入切换到电池电量模拟器。通过给定电池电量进行风光互补系统工作状态切换，避免实训考核过程中长时间等待电池充放电的过程。风光互补发电系统监控软件完成对整个系统的监测与控制（扩展功能）。

系统控制器模块主要由单片机、蓄电池电量检测模块、电池电量模拟器、电池电量值显示等组成。继电器模块主要包括光耦隔离器、继电器和 LED 指示灯。通信接口模块完成 RS-232/485 通信接口协议转换（可扩展功能，具体说明详见项目二）。

（3）系统控制器模块

系统控制器的结构框图如图 3.10.2 所示，主要由主控制器、蓄电池电量检测模块、电池电量模拟器、电池电量值显示和输出控制信号组成。

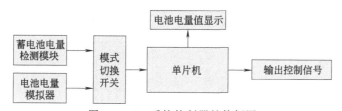

图 3.10.2　系统控制器结构框图

各模块说明：
① 主控制器　由单片机构成；
② 电池电量模拟器　由可调电位器及 AD 转换单元构成，实现电池电量的模拟给定；
③ 电量显示终端　由发光二极管及锁存器构成，显示检测到的电量；
④ 输出控制信号　单片机根据电池电量输出的控制信号；
⑤ 蓄电池电量在线检测模块　检测蓄电池的电量；
⑥ 切换开关 SW1　切换开关打到"0"，则选择了电池电量模拟器；切换开关打到"1"，则选择了蓄电池电量在线检测模块。

图 3.10.3 为系统控制器模块原理图，图 3.10.4 为系统控制器模块实物图，表 3.10.1 为系统控制器模块端口定义。

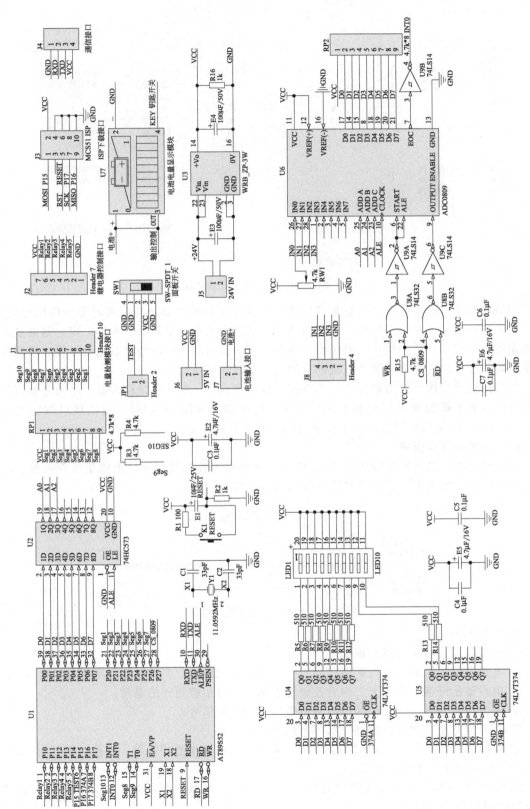

图 3.10.3 系统控制器模块原理图

第三部分 实训项目

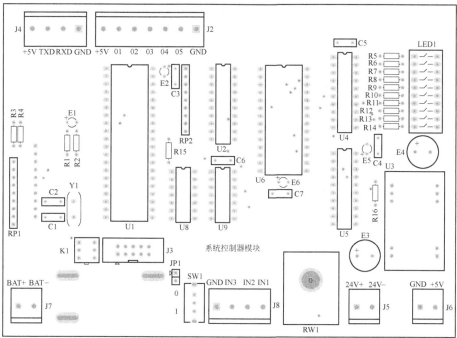

图 3.10.4 系统控制器模块实物图

表 3.10.1 系统控制器模块端口定义

序号	端口定义	说　　明	扩展接口	备注
1	J2:+5V	+5V 电源输出		
2	J2:O1	开关量输出接口 1		
3	J2:O2	开关量输出接口 2		
4	J2:O3	开关量输出接口 3		
5	J2:O4	开关量输出接口 4		
6	J2:O5	开关量输出接口 5		
7	J2:GND	地		

续表

序号	端口定义	说 明	扩展接口	备注
8	J3	89S52 在线下载接口		
9	J4:+5V	+5V 电源输出		
10	J4:TXD	串行口发送端		
11	J4:RXD	串行口接收端		
12	J4:GND	地		
13	J5:24V+	隔离电源 DC/DC 输入 24V 正极		
14	J5:24V−	隔离电源 DC/DC 输入 24V 负极		
15	J6:+5V	+5V 电源输出	√	
16	J6:GND	地	√	
17	J7:BAT+	蓄电池正极		
18	J7:BAT−	蓄电池负极		
19	J8:IN1	模拟量输入 AD0809 通道 1	√	
20	J8:IN2	模拟量输入 AD0809 通道 2	√	
21	J8:IN3	模拟量输入 AD0809 通道 3	√	
22	J8:GND	地	√	
23	JP1	运行程序:短路;下载程序:开路		

(4) 继电器模块

继电器模块由隔离装置及发光二极管、功率继电器构成,发光二极管指示各路输出控制点(即继电器)的状态。图 3.10.5 为继电器模块原理图,图 3.10.6 为继电器模块实物照

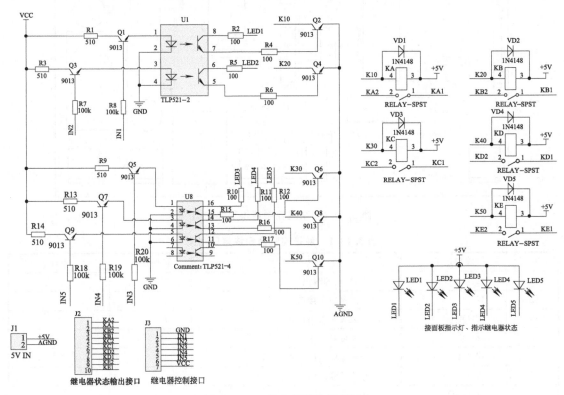

图 3.10.5 继电器模块原理图

片,表 3.10.2 为继电器模块端口定义。

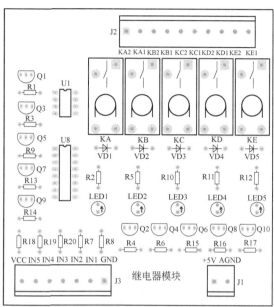

图 3.10.6 继电器模块实物照片

表 3.10.2 继电器模块端口定义

序号	定义	说 明	扩展接口	备注
1	J1:+5V	隔离+5V电源输入正极		
2	J1:AGND	隔离+5V电源输入负极		
3	J2:KA2	KA 继电器常开触点		
4	J2:KA1			
5	J2:KB2	KB 继电器常开触点		
6	J2:KB1			
7	J2:KC2	KC 继电器常开触点		
8	J2:KC1			
9	J2:KD2	KD 继电器常开触点		
10	J2:KD1			
11	J2:KE2	KE 继电器常开触点		
12	J2:KE1			
13	J3:GND	+5V电源输入负极		
14	J3:IN1	KA 继电器控制输入信号		
15	J3:IN2	KB 继电器控制输入信号		
16	J3:IN3	KC 继电器控制输入信号		
17	J3:IN4	KD 继电器控制输入信号		
18	J3:IN5	KE 继电器控制输入信号		
19	J3:VCC	+5V电源输入正极		

图 3.10.7 平板电脑实物图

(5) 平板电脑

平板电脑的实物图如图 3.10.7 所示。

平板电脑的技术指标如下：

① 机箱材质　8mm 硬质氧化铝合金面；

② 处理器　Intel celerom M 900MHz / 2M Cache；

③ 芯片组　Intel 82852 Bridge/Intel Ich 4 South Bridge；

④ I/O 芯片　Winbond W83627DHG-A；

⑤ 系统内存　标配 512M（DDR 200/266）内存；

⑥ BIOS　Award 4 Mb Flash ROM BIOS；

⑦ 扩展　1 个 PCI 插槽（PCI 兼容 Rev.2.2）支持 3 个主 PCI；

⑧ USB　4 个 USB 2.0 接口；

⑨ DIO　1　6 位 GPIO，用于 DI 和 DO；

⑩ 显示器　工业级液晶屏、1024×768 分辨率；

⑪ 产品尺寸　280mm × 210mm × 60mm（D×W×H）；

⑫ MIO　1×VGA、1×S VIDEO-OUT、1×RJ45、4×RS-232、1×AUDIO-OUT、1×CF 卡座、1×PS/2　KEY BOARD / MOUSE、1×4P-DC 座；

⑬ 电源　DC12V 输入。

(6) 接口转换器

CONV485 是一款高性价比的 RS-232 信号与 RS-485 信号互转的接口转换器。该产品直接从设备的串口（如计算机 COM 口）取馈电，无需外接电源。图 3.10.8 为 CONV485 接口图和实物图。

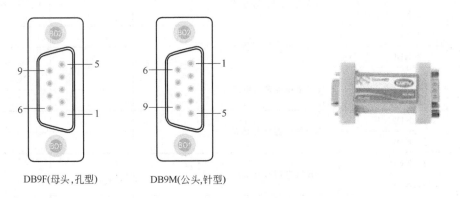

图 3.10.8　CONV485 接口图和实物图

技术指标如下：

接口标准　EIA RS-232、RS485；

RS-232 信号　TXD、RXD、GND；

RS-485 信号　D−、D+、GND；

工作方式　点对点、点对多点的二线半双工；

传输距离　RS-232 端 5m，RS-485 端 1200m；

方向控制　采用数量流向自动控制（DSAC）；
负载能力　支持 32 点；
接口保护　600W 浪涌保护，15kV ESD 保护；
接口形式　RS-232 为 DB9F，RS-485 为 DB9M。

RS-485 接口采用的是 DB9M，RS-232 端口采用 DB9F 接口，如表 3.10.3 所示。

表 3.10.3　接口转换器端口定义

	管脚号	说明		管脚号	说明
	2	RXD		1	485－
RS-232 端(DB9F)	3	TXD	RS-485 端(DB9M)	2	485＋
	5	GND		5	GND

【项目原理及基础知识】

通过能源监控软件将管理单元的硬件（控制器模块、继电器模块和接口转换模块）状态数据进行组态：采集系统组态、数据服务组态、画面组态、网络站点组态、报表开发、脚本编程以及常用的功能参数配置工作。

【项目实施】

(1) 能源监控管理单元的安装与操作

能源监控管理单元包含控制器模块、继电器模块和接口转换模块。模块安装与操作过程如下。

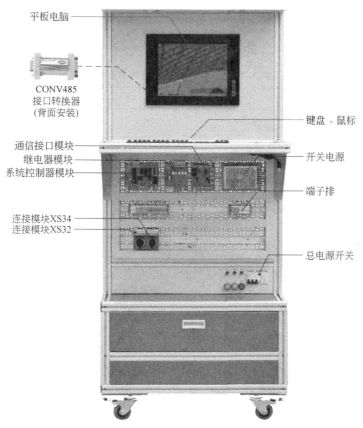

图 3.10.9　能源监控管理单元器件布局图

① 安装　根据如图 3.10.9 所示的器件布局图，将铝导轨、线槽、各功能模块、端子、连接模块等固定到网孔板上，为连线做好准备。

② 接线　端子排端口定义及各部件接线如图 3.10.10、图 3.10.11 和表 3.10.4～表 3.10.10 所示。

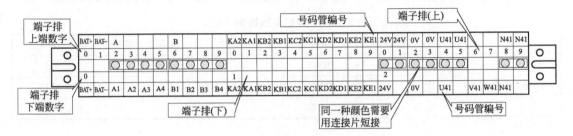

图 3.10.10　端子排端口定义

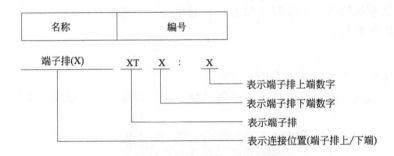

图 3.10.11　端子排编号定义

表 3.10.4　系统控制器模块接线表

序号	起始端位置	结束端位置		号码管编号	线型
	系统控制器模块	名称	编号		
1	J7:BAT+	端子排(上)	XT0:0	BAT+	42 红
2	J7:BAT-		XT0:1	BAT-	42 黑
3	J2:+5V	继电器模块	J3:VCC	VCC	12 蓝
4	J2:01		J3:IN1	402	12 蓝
5	J2:02		J3:IN2	403	12 蓝
6	J2:03		J3:IN3	404	12 蓝
7	J2:04		J3:IN4	405	12 蓝
8	J2:05		J3:IN5	406	12 蓝
9	J2:GND		J3:GND	SGND	12 蓝
10	J5:24V+	端子排(上)	XT2:0	24V	23 红
11	J5:24V-		XT2:2	0V	23 黑
12	J4:TXD	通信接口模块	J9:T1IN	400	12 蓝
13	J4:RXD		J9:R1OUT	401	12 蓝
14	J4:GND		J9:GND	GND	12 蓝

表 3.10.5　继电器模块接线表

序号	起始端位置	结束端位置		号码管编号	线型
	继电器模块	名称	编号		
1	J1：+5V	开关电源	+5V	5V	23 红
2	J1：AGND		COM	0V	23 黑
3	J2：KA2	端子排（上）	XT1：0	KA2	42 红
4	J2：KA1		XT1：1	KA1	42 红
5	J2：KB2		XT1：2	KB2	42 红
6	J2：KB1		XT1：3	KB1	42 红
7	J2：KC2		XT1：4	KC2	42 红
8	J2：KC1		XT1：5	KC1	42 红
9	J2：KD2		XT1：6	KD2	42 红
10	J2：KD1		XT1：7	KD1	42 红
11	J2：KE2		XT1：8	KE2	42 红
12	J2：KE1		XT1：9	KE1	42 红
13	J3：GND	系统控制器模块	J2：GND	SGND	12 蓝
14	J3：IN1		J2：O1	402	12 蓝
15	J3：IN2		J2：O2	403	12 蓝
16	J3：IN3		J2：O3	404	12 蓝
17	J3：IN4		J2：O4	405	12 蓝
18	J3：IN5		J2：O5	406	12 蓝
19	J3：VCC		J2：+5V	VCC	12 蓝

表 3.10.6　通信接口模块接线表

序号	起始端位置	结束端位置		号码管编号	线型
	继电器模块	名称	编号		
1	J1：24V+	端子排（上）	XT2：1	24V	23 红
2	J1：24V-		XT2：3	0V	23 黑
3	J9：T1IN	系统控制器模块	J4：TXD	400	12 蓝
4	J9：R1OUT		J4：RXD	401	12 蓝
5	J9：GND		J4：GND	GND	12 蓝

表 3.10.7　开关电源接线表

序号	起始端位置	结束端位置		号码管编号	线型
	开关电源	名称	编号		
1	L	端子排（上）	XT2：4	U41	23 红
2	N		XT2：8	N41	23 黑
3	EARTH	控制柜	外壳	EARTH	23 黄绿
4	+V	端子排（下）	XT2：0	24V	23 红
5	COM		XT2：2	0V	23 黑
6	+5V	继电器模块	J1：+5V	5V	23 红
7	COM		J1：AGND	0V	23 黑

表 3.10.8　连接模块 XS32 接线表

序号	起始端位置	结束端位置		号码管编号	线型
	连接模块 XS32	名称	编号		
1	A1	端子排（下）	XT0:0	BAT+	42 红
2	A2		XT0:1	BAT−	42 黑
3	A3		XT0:2	A1	12 蓝
4	A4		XT0:6	B1	12 蓝
5	A5		XT0:3	A2	12 蓝
6	A6		XT0:7	B2	12 蓝
7	A7		XT0:4	A3	12 蓝
8	A8		XT0:8	B3	12 蓝
9	A9		XT0:5	A4	12 蓝
10	A10		XT0:9	B4	12 蓝

表 3.10.9　连接模块 XS34 接线表

序号	起始端位置	结束端位置		号码管编号	线型
	连接模块 XS34	名称	编号		
1	B1	端子排（下）	XT1:0	KA2	42 红
2	B2		XT1:1	KA1	42 红
3	B3		XT1:2	KB2	42 红
4	B4		XT1:3	KB1	42 红
5	B5		XT1:4	KC2	42 红
6	B6		XT1:5	KC1	42 红
7	B7		XT1:6	KD2	42 红
8	B8		XT1:7	KD1	42 红
9	B9		XT1:8	KE2	42 红
10	B10		XT1:9	KE1	42 红

表 3.10.10　CONV485 接口转换器接线表

序号	起始端位置	结束端位置		号码管编号	线型
	CONV485 接口转换器	名称	编号		
1	485−	端子排（上）	XT0:2	A	12 蓝
2	485+		XT0:6	B	12 蓝

③ 操作说明

工作状态控制（扩展功能）

将系统控制器模块的"SW1"拨码开关打到"1"端，旋转"RW1"电位器，观察系统控制器模块的 LED 光柱（光柱从下到上每段发光分别表示蓄电池电量为 10%，20%…90%，100%）和继电器模块的发光二极管变化，应能满足下列要求：

　　a. 蓄电池电量显示值＜50%，继电器 KA 吸合，发光二极管 LED1 亮；

b.蓄电池的电量显示值≥50%且＜70%，继电器 KA、KB 吸合，发光二极管 LED1、LED2 亮；

c.蓄电池的电量显示值≥70%且＜90%，继电器 KA、KB、KC 吸合，发光二极管 LED1、LED2、LED3 亮；

d.蓄电池的电量显示值≥90%且＜100%，继电器 KA、KB、KC、KD 吸合，发光二极管 LED1、LED2、LED3、LED4 亮；

e.蓄电池的电量显示值＝100%，继电器全部吸合，发光二极管全亮。

继电器说明（扩展控制功能）

a.继电器 KE 的常开两个触点（KE1、KE2）已经通过航空连接线引出至模拟能源控制单元中。

b.继电器 KC、KD 的常开两个触点（KC1、KC2，KD1、KD2）已经通过航空连接线引出至能源转换储存控制单元中。

c.继电器 KA、KB 的常开两个触点（KA1、KA2，KB1、KB2）已经通过航空连接线引出至并网逆变控制单元中。

用户可以将这些引出到各系统中的继电器常开开关串入各系统的控制回路中，从而可以根据蓄电池的电量控制各系统的工作状态。如当蓄电池电量低时，可将光伏输出、风机输出接入到系统中，对蓄电池进行单充电；当蓄电池电量达到一定程度时，给直流负载供电或启动并网功能；当蓄电池电量充满时，停止充电等。

蓄电池的电量（或者是模拟电池电量）和各输出控制状态之间的关系

当检测到蓄电池的电量＜50%，系统执行单充电操作，仅有风能发电系统和光伏发电系统给蓄电池充电过程。

当检测到蓄电池的电量≥50%且＜70%，系统对蓄电池充电并且给直流负载供电。

当检测到蓄电池的电量≥90%且＜100%，系统对蓄电池充电，给直流负载供电，给并网逆变器供电。

当检测到蓄电池的电量＝100%，系统对蓄电池不充电，给直流负载供电，给并网逆变器供电。

④ 直流电压表、直流电流表、多功能谐波表地址设置

直流电压表、直流电流表地址设置方法

按"SET"键，数码管显示"PASS"，按"→"键切换到"ADDr"后按"SET"键，数码管显示当前地址且闪烁，通过"←"和"→"来改变地址，设置好地址后按"SET"键返回。再按"→"键切换到"bAUD"后按"SET"键，数码管显示当前的波特率且闪烁，通过"←"和"→"来改变波特率，设置好波特率后按"SET"键返回。再通过"←"和"→"切换到"SAVE"后，按"SET"键保存并退出。

a.光伏输出电压表　通信地址（ADDr）：04；波特率（bAUD）：9600。

b.光伏输出电流表　通信地址（ADDr）：01；波特率（bAUD）：9600。

c.蓄电池电压表　通信地址（ADDr）：05；波特率（bAUD）：9600。

d.蓄电池电流表　通信地址（ADDr）：02；波特率（bAUD）：9600。

e.逆变输入电压表　通信地址（ADDr）：06；波特率（bAUD）：9600。

f.逆变输入电流表　通信地址（ADDr）：03；波特率（bAUD）：9600。

多功能谐波表地址设置方法

在主菜单中按"▲"或"▼"，选中"用户设置"再按"确认"进入密码，按"→"设置密码为"0001"，点击"确认"进入，通过"▲"或"▼"选中"通讯设置"，按"确认"

进入,找到"通讯地址",通过"←"和"→"设置通讯地址为:07,"波特率"设置为:9600。

(2) 能源监控管理系统组态设计

每个 SymEnergyV2 制作的工程,都是一个独立的文件夹。尽量不要直接编辑文件夹,以防止工程数据被破坏。

工程管理器(图 3.10.12)提供了工程的创建、删除、已建工程目录的搜索等功能。鼠标双击(或右键菜单),可设置当前的工程。

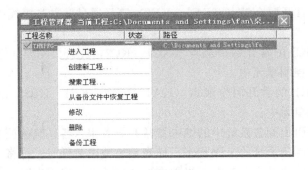

图 3.10.12　工程管理器

① 制作工程的一般步骤

a. 在工程管理器里建立一个新工程(或搜索打开一个工程),进入 SymEnergyV2 集成开发环境(图 3.10.13)。

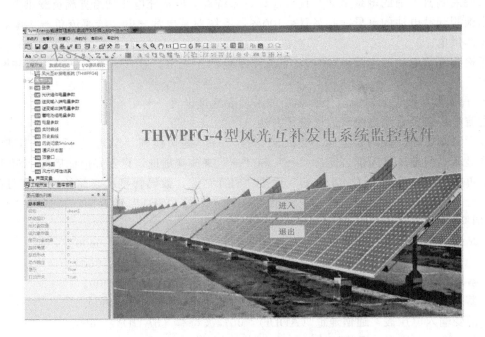

图 3.10.13　SymEnergyV2 集成开发环境

b. 单击 I/O 通讯组态,进入 SymIoServer 组态(图 3.10.14、图 3.10.15),在采集系统下定义采集通道、创建并配置 IED 设备。

图 3.10.14　SymIoServer 组态（一）

IED 智能电子设备是 SymEnergyV2 系统的通讯层的一个对象，与现场的智能装置一一对应。

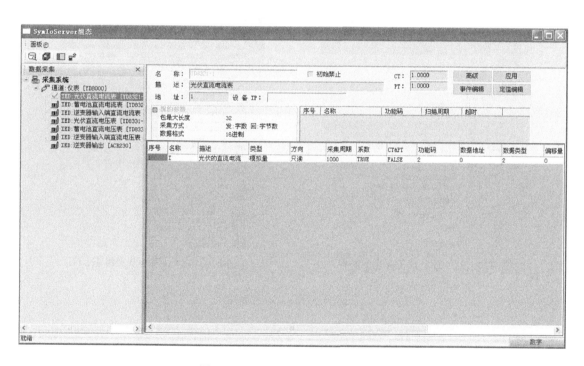

图 3.10.15　SymIoServer 组态（二）

c. 单击工具条上的数据库组态，进入实时数据库组态，定义实时数据库（图 3.10.16）。

实时数据库系统是 SymEnergyV2 平台的核心，几乎所有的 SymEnergyV2 高级功能都是在实时数据库层实现，正确理解实时数据库系统是应用好 SymEnergyV2 系统的关键。

d. 配置实时数据库与 IED 设备变量的映射（图 3.10.17）。

e. 网络节点组态（图 3.10.18）。

f. 配置节点参数（图 3.10.19）。

g. 配置节点运行风格（图 3.10.20）。

h. 规划并创建、配置窗口（图 3.10.21 和图 3.10.22）。

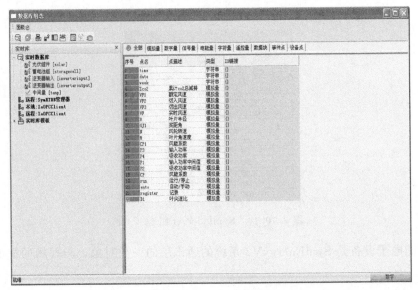

图 3.10.16　实时数据库

图 3.10.17　IED 设备变量的映射

图 3.10.18　网络节点组态

图 3.10.19　配置节点参数

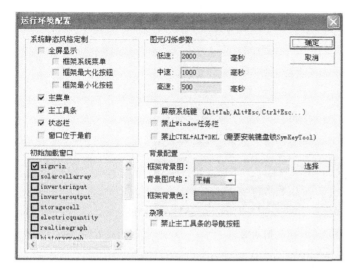

图 3.10.20 配置节点运行风格　　　　图 3.10.21 画面开发

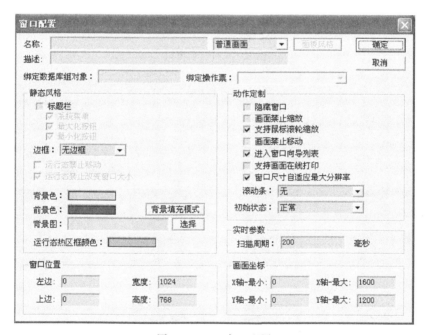

图 3.10.22 窗口配置

i. 画面组态　SymPower 系统提供了三种画面组态模式：
- 类似于组态软件的自由图形绘制和动画配置；
- 基于量值对象的图元绘制；
- 基于设备对象的图元绘制。

② 运行系统（独立服务器）　在开发系统中把工程做完后，双击桌面"独立服务器"图标，运行做完的工程。

(3) 组态设计实例

开启实训设备，连好航空插线和通信线，依次合上实训台的"总电源"和控制柜的"控制电源"开关。打开触摸屏，用工程管理器打开 THWPFG-4 型风光互补发电系统源程序，然后双击桌面"独立服务器"图标，运行工程，出现如图 3.10.23 界面。

图 3.10.23　初始窗口

点击"进入"按钮，进入监控系统之后，出现两个窗口，从上到下依次为"顶窗口"（图 3.10.24）和"通讯状态图"（图 3.10.25）。

| THWPFG-4型风光互补发电系统监控软件 | 系统图 | 通信状态图 | 历史记录 | 历史报表 | 电量参数 | 实时曲线 | 历史曲线 | 风力机特性仿真 | 登录 | 退出 | 9:54:33　2017/2/7　星期二 |

图 3.10.24　顶窗口

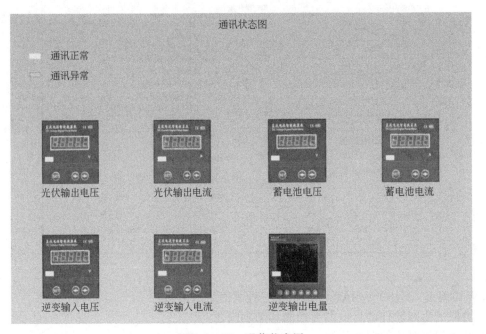

图 3.10.25　通信状态图

"顶窗口"集合了一些主要操作，包括"系统图""通信状态图""历史记录""历史报表""电量参数""实时曲线""历史曲线""风力机特性仿真""登录""退出"。

系统运行时"顶窗口"一直显示，在任何窗口点击"通信状态图"按钮，快速返回"通讯状态图"窗口。此窗口可监测监控系统与直流电压表、直流电流表和交流功率表的通信状态。如果设备通信正常，则各装置的指示图标显示绿色，否则显示红色。

点击"系统图""通信状态图""历史记录""历史报表""电量参数""实时曲线""历史曲

线""风力机特性仿真"按钮,进入各个窗口。点击"登录",出现登录界面(图 3.10.26),进行登录。"退出"按钮用来进行登录状态下的退出系统操作,处于没有登录的状态下则会弹出用户登录对话框(图 3.10.26)。用户名:Admin,密码:(无密码),权限:无限制。

图 3.10.26　用户登录

【项目作业】

① 在工程监控界面绘制监控画面。

② 正确设置组态监控工程各项参数,并采集数据。

参考文献

[1] 杨金焕. 太阳能光伏发电应用技术. 北京：电子工业出版社，2014.
[2] 刘靖. 应用光伏技术. 第 2 版. 北京：化学工业出版社，2016.
[3] GB/T 19565—2017 总辐射表
[4] 陈国呈. PWM 逆变技术与应用. 北京：中国电力出版社，2007.
[5] 徐大平. 风力发电原理. 北京：机械工业出版社，2011.

Bilingual Textbooks of Vocational Education
职业教育双语教材

Installation and Commissioning of Wind-solar Complementary Power Generating System

风光互补发电系统安装与调试

Edited by Song Yao Jie Shen
姚 嵩 沈 洁 主编

Reviewed by Yunmei Li
李云梅 主审

Chemical Industry Press
化学工业出版社

·Beijing·

·北京·

Foreword

In order to expand the cooperation with countries along the BRI routes and implement the requirements for promoting and sharing Tianjin's excellent achievements in vocational education worldwide, vocational education is now playing a pivotal role because of its close relations with the manufacturing industry. According to the cooperation intention of "Luban Workshop", the first batch of Egyptian teachers to be trained in China participated in the Egyptian Luban Workshop EPIP Faculty Training Seminar in Tianjin, and had a one-month teaching training. "Luban Workshop" established in Egypt aims to cultivate urgently-needed technical and skilled personnel for new energy companies and Chinese companies in Egypt and help achieve high-quality resource sharing. In order to serve the teaching of "Luban Workshop" in Egypt in both theory and practice, carry out exchanges and cooperation, enhance the international influence of China's vocational education, innovate the international cooperation mode of tertiary colleges, and share high-quality vocational education resources in China, the research group compiled the textbook entitled *Installation and Commissioning of Wind-solar Complementary Power Generating System*.

This textbook, meticulously arranged, adopts the project-oriented and task-driven concept and builds a curriculum combining the contents of the basic knowledge of photovoltaic power generation technology and wind power generation technology with the practical simulation system training. The book consists of three parts. The first part introduces the wind-solar hybrid power generation technology, mainly focusing on the basic knowledge. The second part is about the wind-solar hybrid power generation system and structure, focusing on the basic parameters of the wind and solar hybrid training platform. The third part of the training project is the key content of the book. In the third part, from project 1 to project 3, the focal points are to enable students to master the theoretical basis of wind-solar hybrid power generation system, theory and practice of solar tracking, volt-ampere characteristic testing technology of components and the method to select appropriate components and batteries. Project 4 to Project 5 mainly dwells on the curve of wind turbine performance and the algorithms of tracking the maximum power of the PV array. Project 6 to Project 7 mainly study the use of off-grid inverters, grid-connected inverter applications, and grid-connected inverter parameter settings and power quality analysis methods. Project 8 to Project 10 is about the application of wind and solar hybrid power generation training system, the running and debugging of photovoltaic power generation system and wind and solar hybrid power generation system and the designing of configuration of energy monitoring management system.

Yao song and Shen Jie supervised the compilation and publication of the textbook and reviewed by Li Yunmei. Yao song, Shen Jie, Li Na, Ma Sining, Sun Yan, Pi Linlin, Wang Xin and Li Liangjun participated in compilation and editing. Yao song arranged the framework, Ma Sining put together the overall compendium, and Shen Jie reviewed the overall framework and all the contents. More specifically, the first and second part is co-complied by Yao Song and Shen Jie. Shen Jie took charge of project 1, Yao song project 2, Ma sining project 3 and project 9, Li Liangjun project 4, Pi Linlin project 5, Sun Yan project 6 and 7, Wang Xin project 8, Li Na project 10.

Owing to our limited knowledge, there must be some omissions and lapses. Your suggestions would be appreciated.

Editor
November, 2019

Contents

Section Ⅰ **Wind-solar Complementary Power Generation Technology** 1

Section Ⅱ **Wind-solar Complementary Power Generation System and Structure** 7

Section Ⅲ **Practical Training Project** 11

Project Ⅰ Installation and Commissioning of Photovoltaic Solar
Tracking System 11

Project Ⅱ Volt-ampere Characteristic Test of PV Module 20

Project Ⅲ Design of System Components and Battery Capacity 29

Project Ⅳ Wind Turbine Characteristics Curve Test 36

Project Ⅴ Verification of Maximum Power Tracking Algorithm
for Photovoltaic Array 48

Project Ⅵ Learn the Principle of Off-grid Inverter and Complete
the System Test 58

Project Ⅶ Test of Grid-connected Inverters and Analysis of Power Quality 65

Project Ⅷ Operation and Debugging of Photovoltaic Power Generation System 88

Project Ⅸ Operation and Debugging of Wind-Solar Complementary
Generating System 93

Project Ⅹ Configuration Design of Energy Monitoring and
Management System 99

Reference 118

Section Ⅰ

Wind-solar Complementary Power Generation Technology

Ⅰ. Introduction of Photovoltaic Power Generation Technology

1. Basic principle of photovoltaic power generation technology

Solar energy is a kind of radiant energy which can be transferred into electric energy only with the help of a converter. The converter that transfers the solar energy into the electric energy is called solar cell component.

The basis of working principle of solar cell is photovoltaic effect of semiconductor p-n junction. The photovoltaic effect refers to an effect that electromotive force and current are generated by changes in the state of charge distribution inside when an object is exposed to the sunlight.

The power generation process of semiconductor solar cell can be summarized as following four aspects:

① Collect the sunlight and other light firstly to ensure that the solar cell surface is exposed to them.

② The solar cell can observe photons with some energy to stimulate non-equilibrium carrier (photogenic charge carrier)—the electron-hole pair. These electron and hole pairs should be sufficient in lifetime and would not disappear before they are separated.

③ In terms of the photogenic charge carrier whose electric symbol is opposed, under the effect of built-in electric field in p-n junction of solar cell, the electron-hole pair is separated with the electron on the one side and the hole on the other side. The opposite electric charges are accumulated on both sides of p-n junction and the photovoltaic electromotive force is generated, namely, the photovoltaic voltage.

④ Take out the electrodes on both sides of the solar cell p-n junction and connect it to the load, then a photo-generated current is passed through the external circuit to obtain a power output. In this way, the solar cell will directly convert solar energy (or other light energy) into electric energy. The monocrystalline silicon solar cell is applied to the training device.

This is the basic process for power generation of p-n junction silicon solar cell, as shown in Figure 1.0.1. If dozens of or hundreds of single solar cells are connected in series and in parallel, the package is a solar cell component, and under the illumination of sunlight, electric energy having a certain power output can be obtained.

2. Introduction of solar cell

(1) The classification of solar cell

Classified according to basic materials: monocrystalline silicon solar cells, polycrystalline silicon solar cells and amorphous silicon solar cells.

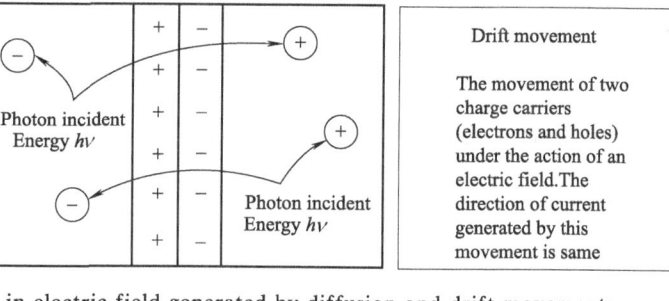

Figure 1. 0. 1   The built-in electric field generated by diffusion and drift movements
caused by combining p-type and n-type materials

Classified according to structure: homo-junction solar cells, hetero-junction solar cells, Schottky junction solar cells, composite junction solar cells and liquid junction solar cells.

Classified according to application: solar cells for space, ground and photovoltaic sensor.

Classified according to using state: flat solar cells, concentrating solar cells and split solar cells.

Classified according to packaging material: rigid packaging solar cells, semi-rigid packaging solar cells and flexible substrate solar cells.

(2) The development mode of photovoltaic power generation

The independent photovoltaic power generation is relative to grid-connected power generation system. It can be called off-grid photovoltaic power generation system. The main purpose of its construction is to solve the power problems in some areas where the power grids cannot be laid. The reliability of power supply in remote areas without electricity is affected by meteorological environment, load and other factors, so the power supply stability is relatively poor. Therefore, off-grid photovoltaic power generation often needs equipment with energy storage and energy management.

The grid-connected photovoltaic power generation system can convert the direct current output from the solar cell array into alternating current which has same range, same frequency and same phase with the power grid voltage, and realize the function of connecting with the power grid and transmitting power to the grid. The power generation system is more flexible, but due to the intermittent nature of the sunshine, the requirements for the performance of the power grid are relatively high. However, the proportion of off-grid photovoltaic power generation in the power grid will gradually increase in the future.

(3) The development trend of photovoltaic power generation

- The off-grid photovoltaic will be widely used in remote areas and nature reserves.
- The places using grid-connected photovoltaic power generation will be more and more flexible and the usage rate of roofs will increase.
- The grid-connected photovoltaic will increase and the absorbing capacity of power grid will increase year by year.

Ⅱ. Introduction of Wind Power Generation Technology

1. Basic principle of wind power generation technology

Wind power generation is a process in which wind energy captured by a wind turbine

Section I   Wind-solar Complementary Power Generation Technology

(fan) is converted into mechanical energy, and the mechanical energy is transmitted to the generator through transmission mechanism such as a main shaft and a gear box, and then the mechanical energy is converted into electric energy by the generator, as shown in Figure 1. 0. 2.

Figure 1. 0. 2   Schematic diagram of the basic principle of wind power generation

Due to the relatively high damage rate of the MW-class wind turbine gearbox, there is a direct-driven wind turbine (without gearbox). The power adjustment methods commonly used in wind turbines are stall adjustment and pitch adjustment.

2. Introduction to wind power generation turbine

(1) Classification of wind power generation turbine

Classified according to wind turbine blade: the paddle type and the pitch type.

Classified according to the speed of the wind turbine: fixed speed type and variable speed type.

Classified according to transmission mechanism: gearbox speed-up type, and direct-driven type.

Classified according to generator: asynchronous type and synchronous type.

Classified according to grid connection: grid-connected type and off-grid type.

(2) Development mode of wind power generation

The direction of onshore wind power generation is low wind speed power generation technology with the main model of 2~5MW large wind turbines and the key is to transmit electricity to the grid.

Offshore wind power generation is mainly used for shallow offshore waters. Large wind turbines of more than 5MW are installed and large-scale wind power generation farms are deployed. The main constraint of this model is the planning and construction cost of wind power generation farms. However, the advantage of offshore wind power generation is obvious, that is, it does not occupy land, and the wind resources at sea are better.

(3) Development trend of wind power generation

① The stall adjustment method will be quickly replaced by pitch adjustment method.

② The constant speed operation mode will be quickly replaced by variable speed operation mode.

③ The scale of the turbines will grow to a larger one.

④ Direct-driven permanent magnet generators and doubly fed Induction generators will jointly develop.

Ⅲ. Introduction of Wind-solar Complementary Power Generation System

The wind-solar complementary power generation system consists of three parts, energy generation, storage and consumption. Solar energy generation and wind power generation are part of the energy generation process which converts solar energy and wind energy with uncertainties into stable energy sources. In order to minimize the imbalance between energy supply and demand due to weather and other factors, batteries are introduced to regulate and balance energy matching, and batteries in the system are used to store energy. Energy consumption refers to various electrical loads, including DC load and AC load. The DC load matching with the working voltage can be directly connected to the circuit, and the DC load which does not match with the working voltage is connected to the circuit after passing through the DC converter. An inverter is required to connect the AC load to the circuit. The wind and solar complementary power generation system is shown in Figure 1. 0. 3. The solar photovoltaic module is used to convert solar energy into direct current electrical energy, and the installation of the solar photovoltaic module adopts a dual-axis tracking system. The PV array has two rotational degrees of freedom that accurately track daylight and ensure that the light illuminates the PV module vertically.

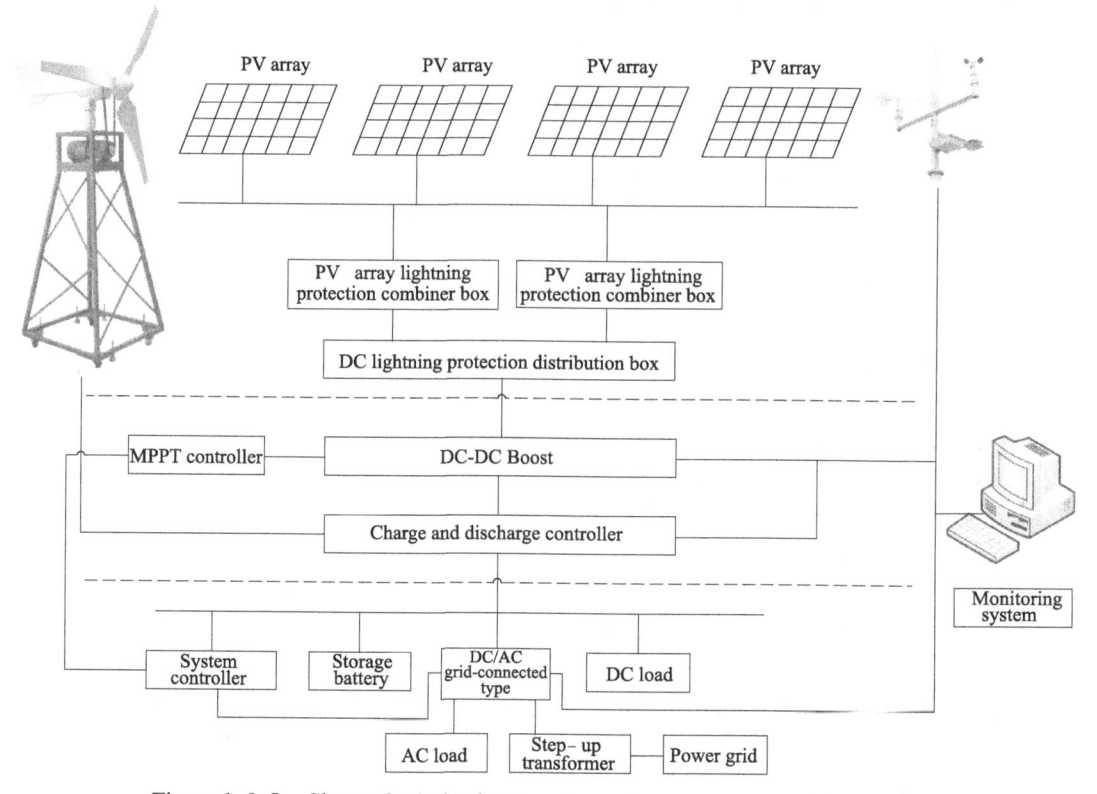

Figure 1. 0. 3   Chart of wind-solar complementary power generation system

In order to reduce the connection between the photovoltaic module and the controller, to facilitate maintenance and to improve reliability, it is generally required to add a DC

Section I Wind-solar Complementary Power Generation Technology

confluence device and a DC distribution box between the PV module and the controller. The PV array can be connected to the MPPT control system after the convergence, and the duty cycle (adjusted equivalent impedance) of the DC-DC Boost converter is adjusted by the MPPT control system to achieve maximum power tracking. It can also be directly connected to a DC load for testing PV characteristics. The output of the DC-DC Boost converter is connected to a intelligent charge and discharge controller for energy storage.

The wind speed is controlled by the variable-frequency drive, and the output power is converted and stored to the battery by the intelligent charge and discharge controller. The battery can be directly loaded with DC load, or converted into AC power through the processing of the inverter device, and then input into the power grid through the transformer to realize grid-connected power generation.

The monitoring system can know the working status of the entire power station in real time, and realize control and display functions, such as light intensity, ambient temperature, photovoltaic battery data monitoring, inverter status monitoring, power generation data monitoring, etc.

IV. Safe Operation Instruction

In order to successfully complete the experimental project of the wind-solar complementary power generation system and ensure the safe, reliable and long-term operation of the equipment during the experiment, the experiment personnel must strictly observe the following safety regulations.

(1) Preparation before the experiment

① Read the instruction manual to familiarize yourself with relevant parts of the system before the experiment.

② Read the system operating instructions and the experiment precautions before the experiment.

③ Read the user manual of the inverter carefully before the experiment to understand the usage of the inverter.

④ Ensure that the power of each system control cabinet is disconnected before the experiment.

⑤ Before the experiment, familiarize yourself with the operation steps of the experiment according to the relevant contents in the experimental instruction book.

(2) Precautions during the experiment

① Strictly follow the correct operating procedures to power on and off the system to avoid damage caused by mis-operation.

② In the process of the operating system, after the battery switch of the energy conversion storage control system is turned on, there is a process of waiting for the self-inspection initialization of the intelligent charge and discharge controller; the red light of the intelligent charge and discharge controller must be extinguished before proceeding to the next step.

③ During the experiment, the temperature of the experimental lighting, the analog light source, the battery board and the surrounding metal firmware will rise under the (long-term) illumination of the light source, so the operator should not touch them directly with the fingers to avoid burns.

④ In the course of the experiment, the place with the "danger" sign has strong electricity, be careful.

⑤ During the experiment, the frequency setting of the analog energy control system inverter should not be too high (not more than 20Hz).

(3) Steps of the experiment

① The preview report is detailed and students should be familiar with the equipment. Before the experiment begins, the instructor should check students' preview reports and ask students to understand the purpose, content and safe operation procedures of the experiment. Only after meeting this requirement can the experiment be allowed.

The instructor should give a detailed introduction to the experimental devices. Students must be familiar with the various devices used in the experiment and clarify the functions and usages of these devices.

② Establish a team and make a reasonable division of labor. Each experiment shall be conducted in groups which consist of 2 to 3 people.

③ Trial run. Before the start of the formal experiment, students should familiarize themself with the operation of the device, then turn on the power according to certain safety practices and observe if the device is normal. If the equipment is abnormal, immediately turn off the power and troubleshoot; if everything is normal, students can start the experiment formally.

④ Be seriously responsible to make the experiment has a beginning and end. After the experiment is completed, the instructor should be asked to check the experimental data. After approval by the instructor, turn off all power sources according to the safe operation steps, and organize the items used in the experiment and put them back in place.

(4) Summary of the experiment

This is the final and most important stage of the experiment. Students should analyze the experimental phenomena and write an experimental report. Each participant must complete an experimental report independently, and the preparation of the experimental report should be serious and realistic.

The experimental report is based on the problems observed in the experiment. After the analysis and research or the analysis and discussion between the members, the summary and the experience of the experience should be written with concise content, identifiable writing and clear conclusion.

The experimental report should include the following:

① Experiment title, major, class, student number, name, name of member in the same group, date of experiment, room temperature, etc.

② Experimental purposes, experimental content and experimental steps.

③ Model, specification, nameplate data and equipment number of the experimental equipment.

④ Collection of experimental data.

⑤ Analyze and summarize the experimental results with theoretical knowledge and draw correct conclusions.

⑥ Analyze and discuss the phenomena and problems encountered in the experiment, write out the experience, and put forward your own suggestions and improvement measures.

⑦ The experimental report should be written on a certain specification of the report paper and be kept clean and tidy.

⑧ Each person should independently complete a report for each experiment and submit it to the instructor on time for review.

Section II

Wind-solar Complementary Power Generation System and Structure

I. System Overview

The platform is an open experimental teaching and research platform for new energy science and engineering and other related majors in colleges and universities. It consists of six parts: simulated light source tracking device, simulated wind energy device, simulated energy control system, energy conversion storage control system, grid-connected inverter control system and energy monitoring management system. The platform is designed with modular structure, and the new energy application system can be designed in many ways. With the integrated application of electronic information, power electronics, automatic control and other technologies, the platform can meet the experimental teaching, engineering design and scientific research innovation of electronic information engineering, electrical engineering and automation, automation, new energy science and engineering and other related majors specialties in colleges and universities.

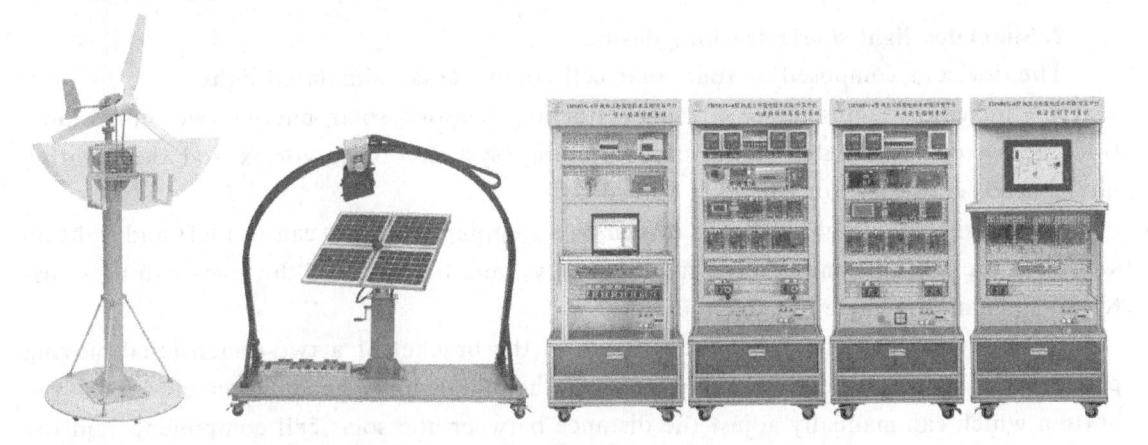

II. System Features

① Modular: adopting industrial standards, it can build different experimental/development systems according to different teaching and research needs with the combination of different modules.

② Openness: parts of software and hardware resources are fully opened to users, and users can add new function modules and carry out inquiry-based and innovative experimental teaching according to their actual needs.

③ Novelty: taking the application of new micro-grid technology as a guide, it combines experiment with design and development.

④ Advantages: the system involves PLC control, frequency converter speed regula-

tion, touch screen, MCS51/PIC MCU, DSP processor and other related technologies with the characteristics of the interdisciplinary and multidisciplinary combination, and it can meet the needs of different levels of users.

III. Technical Performance

① Input power resource: three-phase four-wire AC380V ± 10% 50Hz

② Unit capacity: <3kV · A

③ Dimensions: 1300mm×1100mm×2600mm (simulated wind energy device)

2000mm×1500mm×2000mm (simulated light source tracking device)

800mm×600mm×1880mm (simulated energy control system)

800mm×600mm×1880mm (energy conversion storage control system)

800mm×600mm×1880mm (grid-connected inverter control system)

800mm×600mm×1880mm (energy monitoring management system)

IV. System Structure and Composition

Equipment introduction

1. Experimental/development platform of wind-solar complementary power generation technology

The platform is composed of computer (user-owned), simulated light source tracking device, simulated wind energy device, simulated energy control system, energy conversion storage control system, grid-connected inverter control system and energy monitoring management system.

2. Simulated light source tracking device

The device is composed of four solar cell components, simulated light source (including lamps), solar energy tracking sensor, solar energy two-dimensional tracking system, simulated light source running system, turbo-vortex rod decelerating box, turbo-screw elevator, bracket and so on.

① The simulated light source is driven by a stepping motor. It can run left and right on a circular arc track to simulate the sun trajectory, and the angle of the track can be adjusted to simulate the angle of solar radiation.

② Four solar cell components are fixed on the bracket of a two-dimensional moving platform with a solar energy tracking sensor in the middle and a turbo-screw elevator at the bottom which can manually adjust the distance between the solar cell components and the simulated light source.

③ If it can be combined with simulated energy control system and energy conversion storage control system, the research on photovoltaic automatic tracking (based on sensor or longitude and latitude), solar controller and maximum power tracking (solar energy) and other subjects can be completed.

3. Simulated wind energy device

The device is composed of wind turbine, three-phase variable frequency motor, encoder, drive, fan safety cover and tower.

① The three-phase variable-frequency motor (with an encoder) and the wind turbine are mounted on the tower and are driven by belt. The wind blade rotating surface is equipped

Section II   Wind-solar Complementary Power Generation System and Structure

with a safety cover of semi-circular arc transparent organic glass.

② If simulated energy control system, energy conversion storage control system are combined with it, the research on simulation of wind turbine characteristics, natural wind simulation, maximum power tracking (wind energy), wind-solar complementary controller and DC/DC converter and other subjects can be completed.

4. Simulated energy control system

The system is composed of control screen (power supply, mesh board, tool drawer), programmable logic controller (PLC), programming line, simulation module, frequency converter, touch screen, AC contactor, relay, button, switch and so on.

All the components are installed on the mesh board with open hardware and flexible combination. If solar energy and wind energy devices are combined, the study on photovoltaic automatic tracking (based on sensor or longitude and latitude), simulation of wind turbine characteristics and natural wind simulation and other subjects can be completed.

5. Energy conversion storage control system

The system consists of control screen (power supply, mesh board, tool drawer), PV array confluence module, lightning protection module for DC power supply, DC voltage intelligent digital panel meter, DC current intelligent digital panel meter, disk resistor, circuit breaker, switching power supply, DC voltage current collection module, CPU core module, human-computer interaction module, PWM drive module, communication module, wireless communication module, temperature alarm module, three main circuit modules of DC-DC Boost/Buck/Boost-Buck, battery pack, charge and discharge controller, 51 ISP download unit, PIC programmer and so on.

① The components and modules are all installed on the mesh board. The hardware is open and the combination is flexible. The study on solar controller, maximum power tracking (solar energy, wind energy), wind-solar complementary controller and DC/DC converter and other subjects can be completed with the combination of solar energy and wind energy devices.

② Maximum power tracking processor adopts series 51 which supports on-line download and open hardware. Users can write different MPPT algorithms to realize maximum power tracking and send the adjustment parameters to PWM driver module for adjustment. PWM driving CPU adopts series PIC which receives PWM driving signal of adjusting parameter output isolation, controls main circuit, and realize power regulation.

6. Grid-connected inverter control system

The system consists of DSP core module, interface module, LCD module, keyboard interface module, drive circuit module, boost circuit module, bus voltage sampling module, power grid voltage sampling module, current sampling module, temperature alarm module, communication module, switching power supply, DC load, AC load, DC voltage intelligent digital panel meter, DC current intelligent digital panel meter, inverter output power meter, isolation transformer, off-grid inverter, DSP simulator and so on.

① The components and modules are all installed on the mesh board. The hardware is open and the combination is flexible. The research on PWM control technology, off-grid inverter and grid-connected inverter can be carried out.

② The grid-connected inverter converts DC24V into AC36V and 50Hz, then into

AC220V through a transformer to connect to single-phase mains. The main controller adopts TI fixed point 32-bit TMS320F2812 chip and the output power factor is close to 1. Double closed-loop control is adopted, the inner loop is the current loop, the outer loop is the voltage loop, and the digital phase-locking technology is adopted to synchronize the connected grid.

7. Energy monitoring and management system

The system consists of system controller core module, relay module, communication module, 15-inch industrial tablet computer, keyboard, mouse, configuration software and so on.

The energy monitoring management system can communicate with each control system, the upper computer software can display the running data in real time and can change the running state automatically or manually according to the control requirements and carry out the research on the energy monitoring system.

8. Video teaching software of wind-solar complementary power generation technology

According to the knowledge and skill points of solar power generation, wind power generation and wind-solar complementary power generation systems, the software adopts video teaching method to practice and operate the process of installation, wiring, programming and debugging of the system. The following video teaching contents are included:

① Teaching videos about introduction, installation, wiring and safety of simulated solar energy, simulated wind energy, energy conversion storage control, grid-connected inverter control, energy monitoring management systems/devices.

② Teaching videos of solar automatic tracking, maximum power tracking and grid-connected inverter.

③ Teaching videos about using and programming of PLC, frequency converter, touch scree, MCS51/PIC MCU, DSP processor, configuration software.

9. 3D simulation training software of wind-solar complementary power generation system

The software includes simulated wind, simulated sunlight, wind power generation farm, wind power generator, various types of solar energy battery modules (monocrystalline silicon, poly-crystalline silicon, amorphous silicon), support (fixed, single-axis, double-axis tracking), storage battery pack, wind-solar complementary controller, inverter (off-grid, grid-connected) and AC and DC load (traffic lamp, street lamp, LED screen, water pump) simulation model suitable for various applications. Software shows the structure and working principle of each part with vivid and intuitive simulation animation and 3D model.

It can simulate various power generation systems (such as independent photovoltaic power generation system, independent wind power generation system, wind-solar complementary power generation system, off-grid power generation system, grid-connected power generation system, etc.) and application system examples (such as solar energy traffic lamp, solar energy street lamp, solar energy LED screen, solar energy water pump, wind-solar complementary street lamp, wind-solar complementary monitoring etc.).

Wind farms simulate a variety of real scenes and understand the layout of wind farms and the operation of large-scale wind turbines through roaming and flight mode. 3D models of wind turbine mechanical components are realistic, which can support rotation in any direction and 360-degree and all-round display of the operating state.

Section III

Practical Training Project

Project I Installation and Commissioning of Photovoltaic Solar Tracking System

【Project Description】

Learn the basics of an simulated light source tracking control unit through training platform of THWPFG-4 wind-solar complementary power generation system and use the simulated energy control unit to keep the solar panel always follow the simulated light source.

【Ability goals】

① To learn the programming method of Mitsubishi PLC.

② To ensure that the solar panel can complete the sun-chasing motion through the programming control level and pitching motion mechanism of the Mitsubishi PLC.

【Project Environment】

In order to complete the training task, we should refer to the training platform equipment manual of THWPFG-4 wind-solar complementary power generation system, learn the electrical schematics of the simulated light source tracking control unit, understand the electrical indexes and interface description of the 3ND583 stepping motor driver, the operating principle of the M542 stepping motor, the technical parameters and wiring principles of the solar energy simulated sun tracking sensor, the technical parameters of the stimulated light source and method of wiring in the junction box of solar cell component.

The simulated light source tracking control unit is composed of the solar cell component, the simulated light source, the solar energy stimulated sun tracking sensor, the solar panel two-dimensional motion mechanism, the stepping motor, the stepping motor driver, the deceleration box, the Mitsubishi programmable controller, the button and the relay. The circuit diagram of the stimulated light source tracking control unit is shown in Figure 3.1.1.

The PLC control light of the simulated energy control unit is used to simulate the trajectory of the sun rising east and falling west and the incidence angle of the sun, and the simulated sun tracking sensor on the solar panel collects the illumination information and position information of the simulated light source to control the two-dimensional motion mechanism, so that the solar panel always follows the simulated light source.

Technical parameters are as follows:

Installation and commissioning of ptotovoltaic solar tracking system

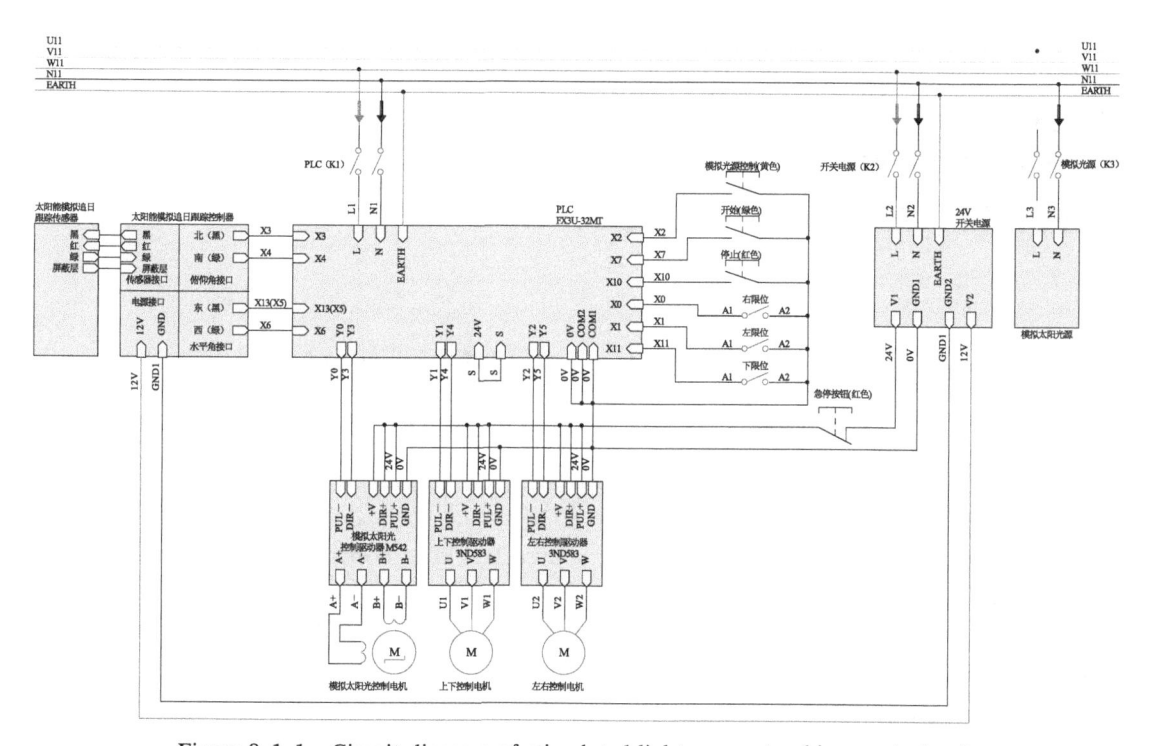

Figure 3.1.1　Circuit diagram of stimulated light source tracking control unit

① Specifications of solar cells：20W/18V×4；

② Simulated light source power：1000W；

③ Tracking mode：double axis，tilting 180°，rotating 360°；

④ Tracking accuracy：＜±1.5°；

⑤ Operating voltage：DC24V；

⑥ Dimensions：2000mm×1200mm×2800mm。

Composition of simulated light source tracking control unit is as follows.

(1) Mitsubishi programmable controller

The input and output end definitions of Mitsubishi programmable controller are shown in Table 3.1.1 and Table 3.1.2.

Table 3.1.1　PLC input end definition

Type	Terminal function	Mitsubishi host port
Input	Right limit switch	X0
	Left limit switch	X1
	Stimulated light source control button	X2
	North sensor signal	X3
	South sensor signal	X4
	East sensor signal	X13
	West sensor signal	X6
	Start button	X7
	Stop button	X10
	Lower limit switch	X11

Section Ⅲ Practical Training Project

Table 3.1.2　PLC output end definition

Type	Terminal function	Mitsubishi host port
Output	Simulated sunlight control driver PUL −	Y0
	Up and down control driver PUL −	Y1
	Left and right control driver PUL −	Y2
	Stimulated sunlight control driver DIR −	Y3
	Up and down control driver DIR −	Y4
	Left and right control driver DIR −	Y5

Note: One end of the pitch angle interface and the horizontal angle interface of the solar tracking sensor is connected to the 0V port, and the other end is connected to the input ports of the host.

(2) 3ND583 stepping motor driver

3ND583 adopts high subdivision three-phase stepping driver designed by precision current control technology, which is suitable for driving three-phase stepper motors with frame size 57~86. The electrical indexes of the driver, is shown in Table 3.1.3.

Table 3.1.3　Electrical indexes

Explanatory note	Minimum value	Typical value	Maximum value	Unit
Output current	2.1	—	8.3(mean 5.9)	A
Input supply voltage	18	36	50	V DC
Control signal input current	7		16	mA
Stepped pulse frequency	0		400	kHz
Pulse low-level time	1.2			μs
Insulation resistance	500			MΩ

Interface description of 3ND583 stepping motor driver is shown in Table 3.1.4 and Table 3.1.5.

Table 3.1.4　P1 port (control signal interface) description

Nomenclature	Function
PUL + (+5V) PUL − (PUL)	Pulse control signal: pulse rising edge is effective; PUL-high level is 4~5V, low level is 0~0.5V. In order to respond reliably to pulse signals, the pulse width should be greater than 1.2μs. If it is used, the string resistance is required
DIR + (+5V) DIR − (DIR)	Direction signal: high/low level signal. in order to ensure reliable direction change of motor, direction signal should be established before pulse signal at least 5μs. The direction of initial operation of the motor is related to the connection of the motor. Changing any two lines of U, V and W of the three-phase winding can change the direction of the initial operation of the motor. DIR-high level is 4~5V and low level is 0~0.5V
ENA + (+5V) ENA − (ENA)	Enabling signal: this input signal is used to enable or disable. When ENA + is connected with +5V, ENA − is connected with low level(or internal optical coupling conduction), the driver will cut off the current of each phase of the motor so that the motor is in free state and then the stepping pulse is not responded. When this function is not required, the signal end can be suspended

Note: When using 24V power resource, it is necessary to string resistance R into the signal end of the controller, as shown in Figure 3.1.7 PLC control schematic diagram (in the latter P17). R is 2kΩ, greater than 1/8W resistance.

13

Table 3.1.5　P2 port（power interface）description

Nomenclature	Function
GND(GND)	DC power supply ground
+ V	DC power supply positive, any value is available from + 18~ + 50V, but the recommended value is about + 36V DC
U	U-phase of the three-phase motor
V	V-phase of the three-phase motor
W	W-phase of the three-phase motor

Protective function of driver

① Under voltage protection　when the DC power supply voltage + V is lower than 18V, the green light of the driver is off and the red light flashes and then it comes into the under voltage protection state. If the input voltage continues to drop to 16V, the red and green lights will be extinguished. When the input voltage recovers to 20V, the drive will automatically reset to normal working condition.

② Over-voltage protection　when the DC power supply voltage + V exceeds 51V DC, the protection circuit action will carry out, the power supply indicator light turns red and the protection function starts.

③ Over-current and short-circuit protection　when the motor connection winding occurs a short circuit or the motor itself damages, the protection circuit action will carry out, the power indicator light turns red and the protection function starts. When the over-voltage, over-current or short-circuit protection function starts, the motor shaft loses self-locking force and the power indicator light turns red. If you want it back to normal work, you should make sure the above faults are eliminated, then turn on the power supply, when the power indicator light turns green, the motor shaft is locked, the driver is back to normal.

（3）M542 stepping driver

M542 driver has the characteristics of low motor heating, low running noise and running stability. It mainly drives 42 and 57 two-phase hybrid stepping motor. The recommended working voltage range is 24~36V DC, the voltage is not more than 50V DC but not less than 20V DC. The using instructions of the driver are shown in Table 3.1.6.

Table 3.1.6　Driver using instructions

Driver function	Operating instructions
Signal interface	PUL + and PUL − are the positive and negative ends of the control pulse signal. DIR + and DIR − are the positive and negative ends of direction signal. ENA + and ENA − are the positive and negative ends of the enabling signal
Motor interface	A + and A − are connected to the positive and negative ends of A-phase winding of the stepping motor; B + and B − are connected to the positive and negative ends of the B-phase winding of the stepping motor. When the A-phase winding and B-phase winding are exchanged, the motor can rotate in the opposite direction
Power supply interface	Using DC power supply, the working voltage range is 24~36V, the power supply power is more than 100W and the voltage is not more than 50V DC and not less than 20V DC
Indicator lamp	There are two lights, red and green, in the right in the driver and the green light is the power indicator light, when the driver is on, the green light is always on; the red light is the fault indicator light. When there is over-voltage or over-current fault, the fault light is always on. After the fault is cleared, the red light goes out. When the driver fails, only turning it on and enabling it again can clean the fault
Installation instructions	The driver can be installed either horizontally or vertically. Attach it to the metal cabinet for heat dissipation when installing

Section Ⅲ   Practical Training Project

(4) Solar energy stimulated sun tracking sensor

The sensor and the controller only detect the position deviation between the sensor and the simulated light source and transform the position signal into four directional switching signals. When the sensor is aligned with the simulated light source, none of the four signals is output. A cloudy-sunny day adjustment potentiometer is used to detect the strength of sunlight or the signal of day and night. Adjust the potentiometer clockwise till hearing the relay sucks and then adjust it counterclockwise till hearing the relay opens. When blocking the front end of the sensor with hands, the relay will suck.

Technical parameters

① System power supply: DC12V;

② Tracking accuracy: 1°;

③ Signal output mode: passive contact;

④ Number of signals: 4/5;

⑤ Output signal definition: left, right, up, down/cloudy and sunny (day/night).

The controller is composed of two parts: a sunlight sensor (RY-CGQ-1-S) and a controller (RY-KZQ-D).

A sensor and a controller are shown in Figure 3.1.2 and Figure 3.1.3.

Figure 3.1.2   Sunlight sensor (RY-CGQ-1-S)

Figure 3.1.3   Controller (RY-KZQ-D)

The definitions of the control box outgoing line are shown in Table 3.1.7.

Table 3.1.7   Control box outgoing line definition

Nomenclature	Definition	Remark
Sensor interface	Line-1	The sensor interface lines should correspond to the colors one by one and the shield layer should be connected
Power supply interface	Line-2	
Pitch angle interface	Line-3	
Horizontal angle interface	Line-4	

Power supply interface	Line-2
Brown(red)	DC12V+
Blue(black)	DC12V−
Yellow	NC
Green	NC

Pitch angle interface	Line-3
Red(0V)	Northward movement
Black(X3)	
Yellow(0V)	Southward movement
Green(X4)	

Horizontal angle interface	Line-4
Red(0V)	Eastward movement
Black(X13)	
Yellow(0V)	Westward movement
Green(X6)	

Note: The pitch angle interface and the horizontal angle interface are connected to 0V at one end and the host input port at the other end.

(5) Simulated light source

The simulated light source uses ultra-bright halogen tungsten lamp which is used to simulate sunlight illuminating solar modules to generate electric energy. Simulated light source is shown in Figure 3.1.4.

Figure 3.1.4   Simulated light source

Technical parameters of simulated light source are shown in Table 3.1.8 below.

Table 3.1.8   Technical parameters of simulated light source

Model	QVF137	Lamp base	R7s
Voltage/frequency	220V/50Hz	Weight/kg	1.75
Light source	PLUS PRO L 1000W R7s	Windward area/mm^2	0.054

Characteristics of simulated light source

① Anode alumina reflector provides effective wide light distribution, and it is more suitable for wide-range lighting applications.

② Water-proof and dust-proof grade reaches IP54.

③ High-pressure cast aluminum lamp body, anti-corrosion surface coating.

④ Hinged tempered glass, it is easy to operate, and no maintenance is required.

(6) Solar cell component

Solar panels are the core part of solar power generation systems and the most valuable part of solar power generation systems. The role is to convert the radiant power of the sun into electric energy, or to send it to the batteries for storage, or to drive the load to work. The solar cell component is shown in Figure 3.1.5.

When more solar batteries are in series to form a branch of a battery array or a battery

Section Ⅲ　Practical Training Project

matrix，it is necessary to parallel 1 (or 2~3) diode inversely at the positive and negative output ends of each panel，which is called a bypass diode. The junction box and the bypass diode are shown in Figure 3.1.6.

Figure 3.1.5　Solar cell component

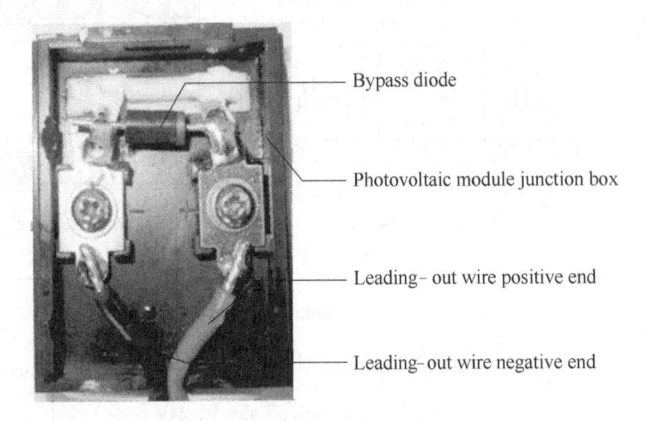

Figure 3.1.6　Junction box and bypass diode

【Project Principle and Basic Knowledge】

(1) Light source simulated tracking device

The system is composed of a light source simulated tracking device and a light source simulated tracking control system. The system consists of solar cell component，simulated solar lamp，solar simulated sun tracking sensor，solar panel two-dimensional motion mechanism，stepping motor，stepping motor driver，deceleration box，Mitsubishi programmable controller，button and relay.

Main parameters of solar panel component：Nominal power 20W；Operating voltage 17.5 V；Operating current 1.14 A；Open circuit voltage 22.0 V；Short circuit current 1.23A。

(2) PLC control schematic (Figure 3.1.7)

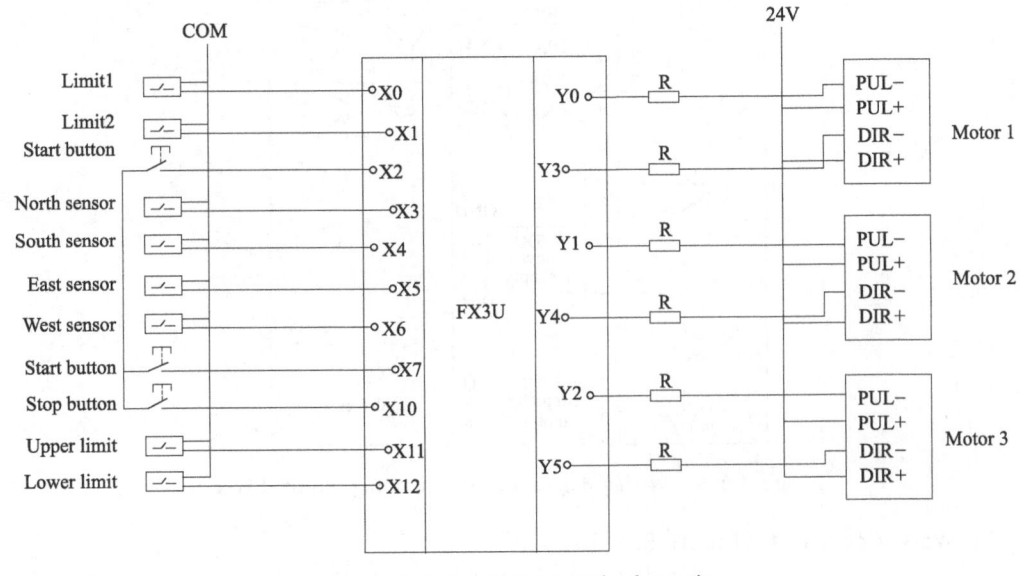

Figure 3.1.7　PLC control schematic

17

(3) Wiring diagram of the control cabinet (Figure 3. 1. 8)

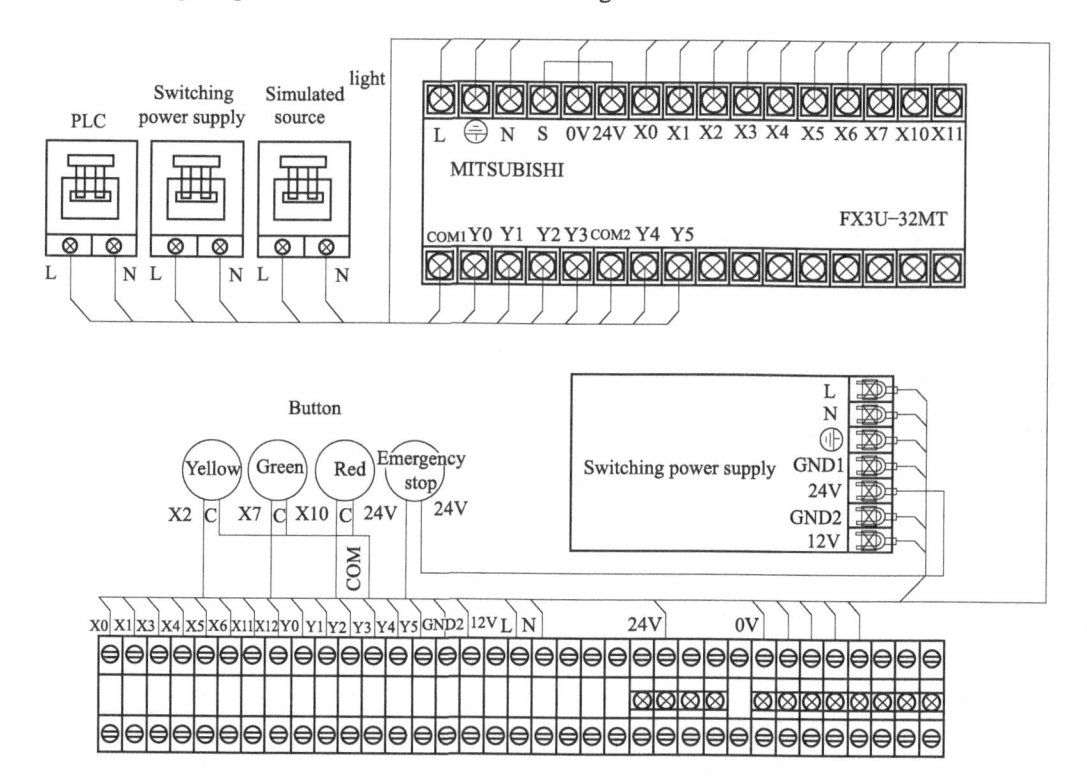

Figure 3. 1. 8   Wiring diagram of the control cabinet

(4) Wiring diagram of the stepping motor driver (Figure 3. 1. 9)

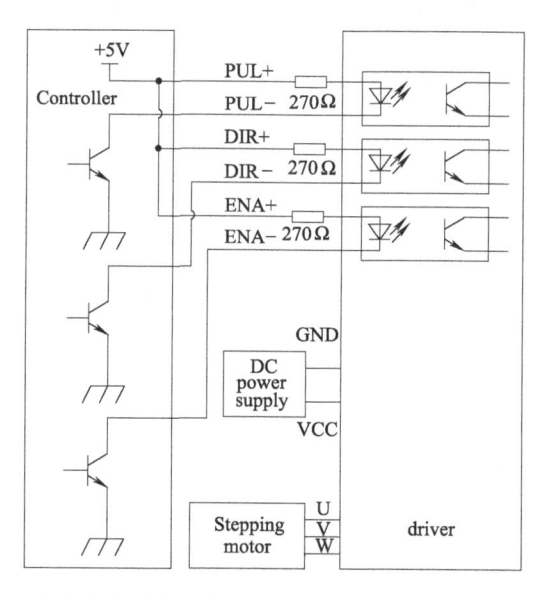

Figure 3. 1. 9   Wiring diagram of the stepping motor driver

(5) Work flow chart (Figure 3. 1. 10)

(6) Signal definition (Table 3. 1. 1 and Table 3. 1. 2)

Section Ⅲ　Practical Training Project

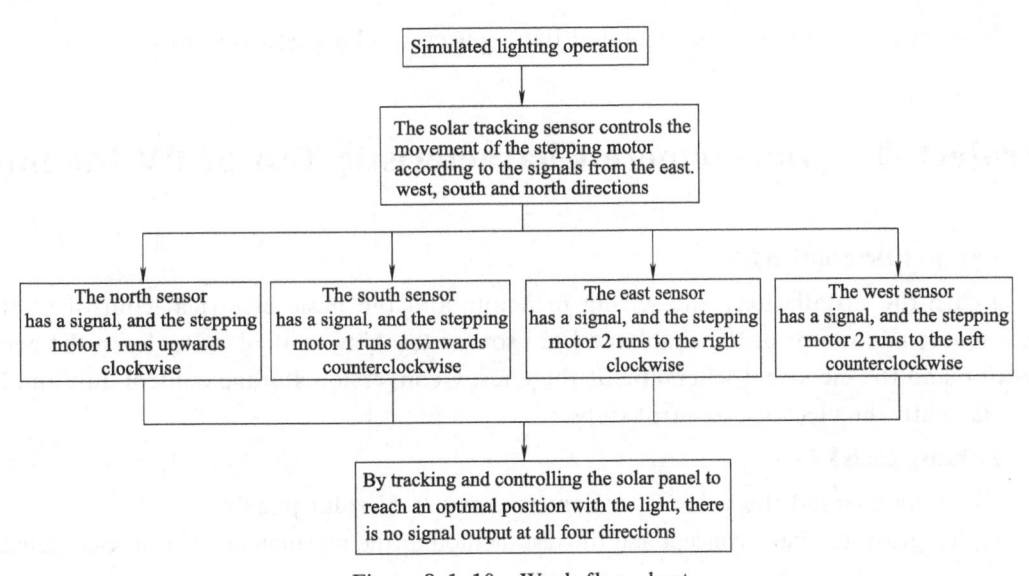

Figure 3.1.10　Work flow chart

【Project Implementation】

① Turn on the PLC switch, the power switch and the simulated light source switch of the "simulated energy control system" in turn.

② Connect the PLC to the computer by using the PLC download line, then download the "solar panel sun tracking" sample program.

③ Turn the "RUN/STOP" switch of the PLC to the "RUN" state, and the system will run the sample program.

Solar panel tracking system

④ Press the "start" button (green button) of the sun tracking system control switch to prepare the stepping motor.

⑤ Press the "control" button (yellow button) to start the motion of the simulated light source; the sensor on the solar panel controls the stepping motor according to the light sensing signal, which causes the solar panel move synchronously with the stimulated light source. After the simulated light source moves to the left limit baffle and returns to the right limit baffle along the original path, one cycle completes. The simulated light source will repeat the periodic motion before the stop button is pressed (The "control" button has a self-locking function. After it is pressed, the simulated light source can move).

⑥ Press the "stop" button (red button) and all stepping motors will stop. After pressing the "start" button, they will continue running.

⑦ Press the "emergency stop" button to cut off the 24V power supply of the stepping motor driver, and the system continues operating after releasing the button.

⑧ After the experiment finishes, press the "stop" button (red button) to stop all stepping motors. Then turn off the PLC switch, the switch power switch, the simulated light source switch and the main power switch of the simulated energy control system in turn. If the subsequent experiment will be carried out, the main power switch can not be turned off.

【Project Exercise】

① Draw the hardware control chart, the wiring diagram and the PLC program flow chart of the solar panel sun tracking system.

② According to the PLC program flow chart, write the solar panel sun tracking control program.

19

③ What is the core of the simulated light source tracking control unit?

Project Ⅱ   Volt-ampere Characteristic Test of PV Module

【Project Description】

Learn the installation, the wiring principle and the basic electrical control of the light source and motor of the simulated light source tracking control unit, learn the composition and the electrical schematic of the energy conversion storage control unit and be familiar with the electrical control steps.

【Ability goals】

① To understand the volt-ampere characteristics of solar panels.

② To grasp the basic concept and the test method of the maximum power of solar panels.

【Project Environment】

To complete the training task, we should refer to the training platform equipment instruction manual of THWPFG-4 wind-solar complementary power generation system.

Volt-ampere characteristic test of PV module

The energy conversion storage control unit is mainly composed of PV array junction module, DC power supply lightning preventer module, DC voltmeter, ammeter, circuit breaker, switching power supply, MPPT controller, battery pack and intelligent charge and discharge controller. It mainly completes the functions of PV confluence, lightning protection, electric quantity measurement, maximum power tracking, energy storage and battery management. The electrical circuit diagram of the energy conversion storage control unit is shown in Figure 3.2.1.

Function description of each module

① Photovoltaic array junction module it is used to input the multi-channel output cables of the solar cell component square array together and connect them by group. It not only makes the wiring orderly, but also makes it easy to check and maintain in group.

② DC lightning preventer module lightning preventer is a protective device which is used to prevent lightning surge voltage from intruding into solar cell array, AC inverter, AC load or power grid.

③ DC voltage, ammeter monitoring electricity quantity parameters, such as photovoltaic output voltage, photovoltaic output current and so on.

④ Circuit breaker it mainly completes the disconnection and access of each branch.

⑤ Connection module it can connect the control object and the controller, and it is convenient to plug and pull the connecting wire.

⑥ Switching power supply it provides stable DC power supply for circuit and equipment.

⑦ MPPT controller it mainly completes PV and fan maximum power tracking (MPPT) algorithm.

⑧ Battery pack it is additional batteries in the grid-connected system, it can supply electricity under the condition of no wind and no light, and it can also expand the scope of use of the system, such as buffering, storing electricity, regulating power peak, in case

20

Section Ⅲ   Practical Training Project

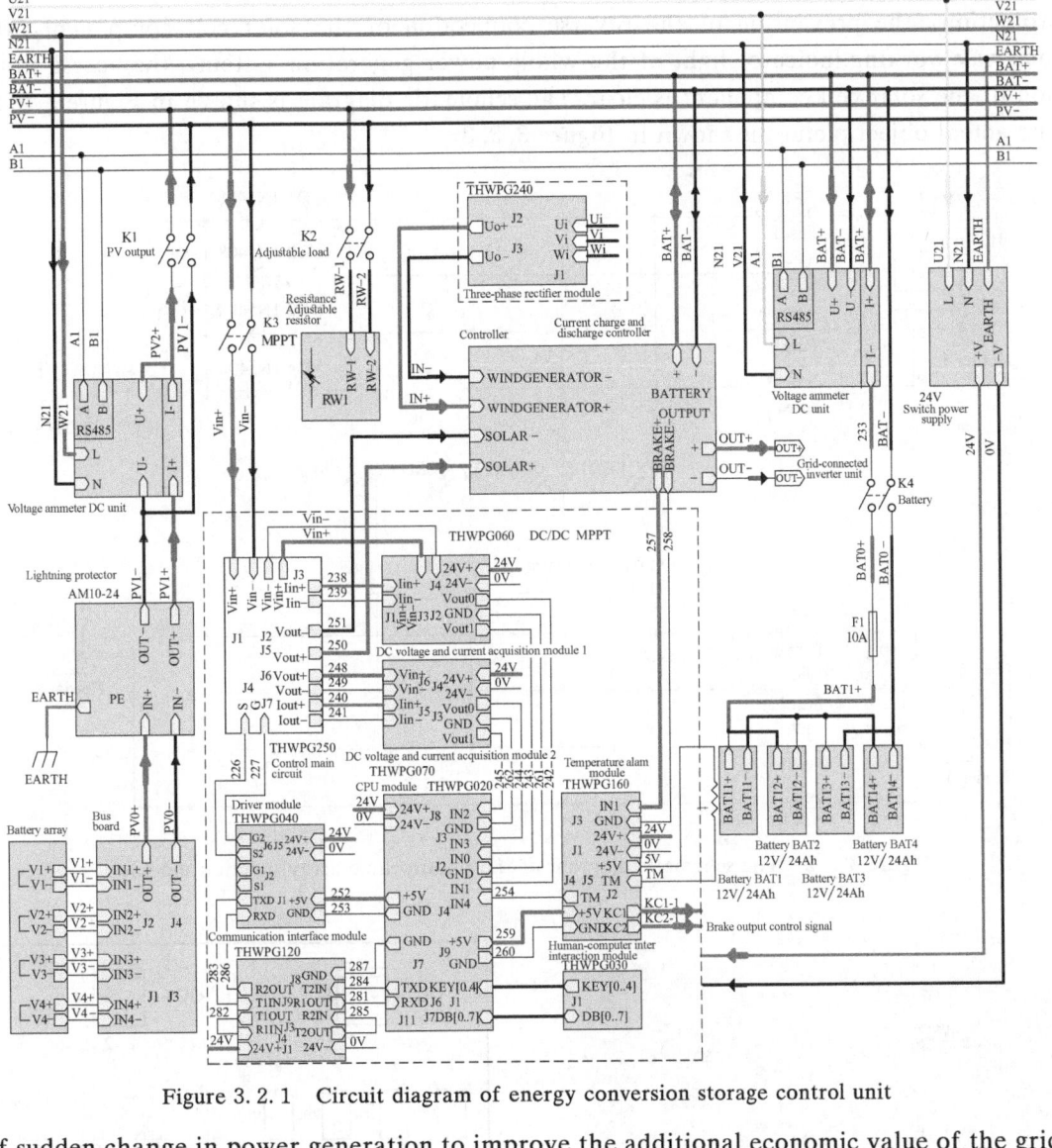

Figure 3. 2. 1   Circuit diagram of energy conversion storage control unit

of sudden change in power generation to improve the additional economic value of the grid-connected system.

⑨ Intelligent charge and discharge controller   according to the voltage of battery, it adjusts the charging state and current and prevent the battery from overcharging or over-discharging to prolong the battery life.

In the training, we should focus on the electrical connection between the output and the adjustable load of the PV module, and ensure that the battery is disconnected from the circuit during the experiment.

(1) Photovoltaic array junction module

The photovoltaic array junction module connects four solar cell components together and then output them.

The multi-channel output cables of solar cell components are connected centralizedly in groups, which not only makes the wiring in good order, but also facilitates the inspection and maintenance in groups. When the solar battery array has local faults, they can be

21

partially separated and repaired without affecting the continuous work, the over-current protection, the prevention of the reverse connection of the solar cell component, the four-way working indicator light of the whole power generation system, the access situation of the solar cell component is clear. The schematic diagram is shown in Figure 3.2.2. The actual object picture is shown in Figure 3.2.3

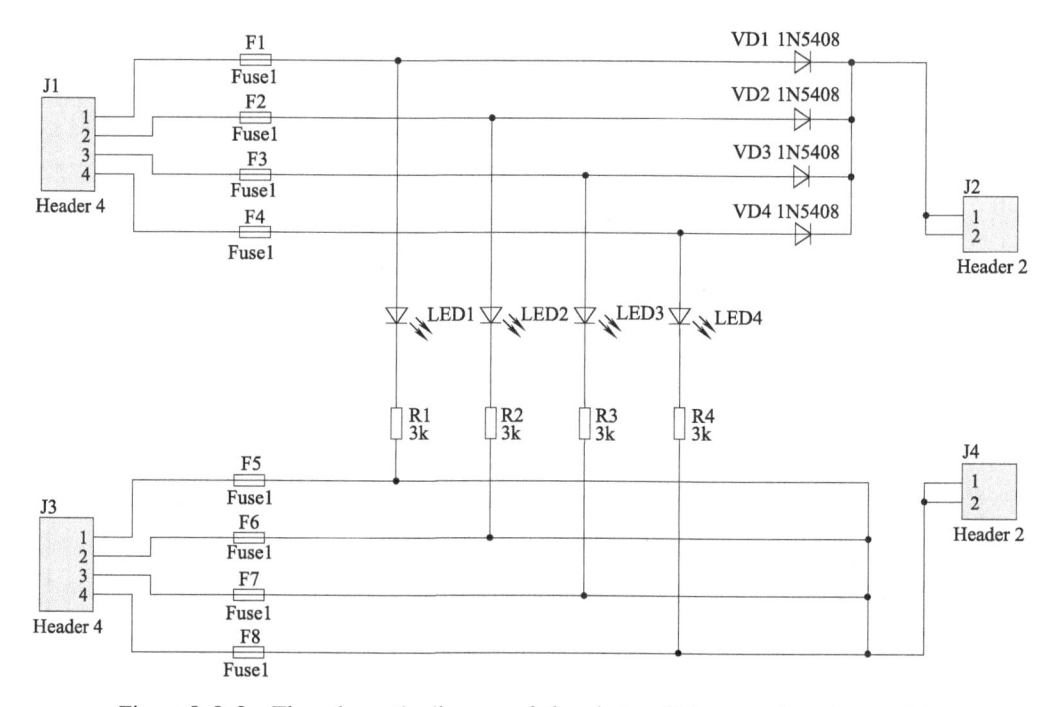

Figure 3.2.2　The schematic diagram of the photovoltaic array junction module

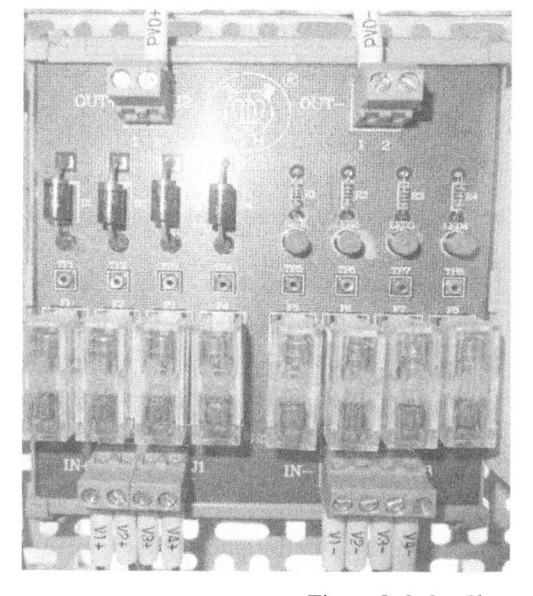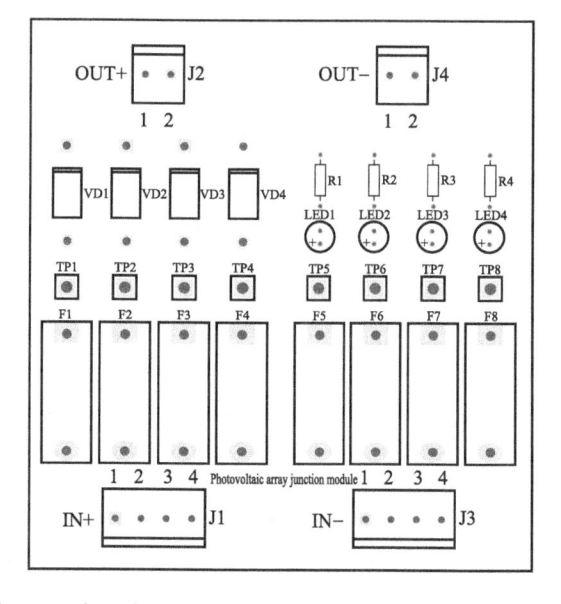

Figure 3.2.3　Photovoltaic array junction module

The port definitions are shown in Table 3.2.1.

Section Ⅲ   Practical Training Project

Table 3. 2. 1   Port definitions

S/N	Definition	Explanatory note
1	J1:1(IN+)	DC positive input of path 1
2	J3:1(IN−)	DC negative input of path 1
3	J1:2(IN+)	Direct current positive input of path 2
4	J3:2(IN−)	DC negative input of path 2
5	J1:3(IN+)	DC positive input of path 3
6	J3:3(IN−)	DC negative input of path 3
7	J1:4(IN+)	DC positive input of path 4
8	J3:4(IN−)	DC negative input of path 4
9	J2:1/2(OUT+)	DC converging positive output
10	J4:1/2(OUT−)	DC converging negative output

(2) DC lightning preventer module

Lightning preventer is a protective device which is used to prevent lightning surge voltage from intruding into solar battery array, AC inverter, AC load or power grid. Main function: to protect the solar battery array, lightning-proof parts should be installed in each component string; it adopts the technology of temperature-controlled circuit break and built-in over-current protection circuit to avoid the risk of fire caused by heating of the lightning preventer completely. Working status and failure status display clearly and intuitively. The schematic diagram is shown in Figure 3. 2. 4. The technical indicators are shown in Table 3. 2. 2. The actual object picture is shown in Figure 3. 2. 5. The port definitions are shown in Table 3. 2. 3.

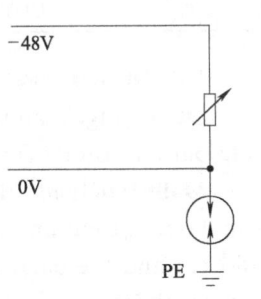

Figure 3. 2. 4   Fundamental diagram of lightning preventer module of DC power supply

Table 3. 2. 2   Main technical parameters

Product model	AM10∼24
Nominal operating voltage U_n	24V DC
Maximum continuous working voltage U_c	36V DC
Nominal discharge current(8/20μs)	5kA
Maximum flow capacity I_{max}(8/20μs)	10kA
Protection level U_p(at I_n)	≤150V
Response time	≤25ns
IP protection level	IP20
Flame retardant grade, in accordance with UL94	V0
Area of access wire	+, − and 0 wire ≥6mm², and the earth wire ≥10mm²
Overall dimension	Single: 90mm×18mm×69mm; Double: 90mm×36mm×69mm
Working environment	Temperature −40∼+85 ℃, relative humidity ≤95%(25℃), height ≤3km

23

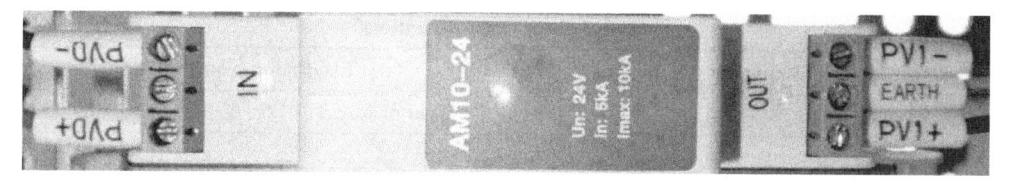

Figure 3.2.5   Lightning preventer module of DC power supply

Table 3.2.3   Port definitions

S/N	Definition	Explanatory note
1	IN+	DC power supply positive input
2	PE	Earth wire(lightning-proof ground)(earth)
3	IN−	DC power supply negative input
4	OUT+	DC power supply positive output
5	PE	Earth wire(lightning-proof ground)(earth)
6	OUT−	DC power supply negative output

(3) DC unit module

DC voltage, ammeter: monitoring electricity quantity parameters, such as photovoltaic output voltage, photovoltaic output current, etc.

Main functions DC voltage (current) measurement, display; RS485, serial output, simulated output and upper and lower limit alarm, control (contact output); It can set the address and the baud rate, and display value, upper and lower limit alarm valve value and return difference by hand or upper computer.

Technical features it adopts modular design, can choose a single display according to the needs of users and display plus other functions, and adopts the latest PIC chip with strong anti-interference ability.

General description the installation structure is disk installation, wiring on the back; the shell material is flame-retardant plastic; the isolation acts as the mutual isolation of power supply, input and output.

Performance index

① Maximum indication: ±199999

② Display resolution: 0.001

③ Maximum input range: voltage   DC 0~600V, current   DC 0~5A

④ Input mode: single end input

⑤ Accuracy level: 0.5

⑥ Absorption power: <1.2V・A

⑦ Measuring speed: about 5 times per second

⑧ Output load capacity: ≥300Ω

⑨ Power supply: AC220V±20%, 50/60Hz

⑩ Power frequency voltage withstand: among power supply/input/output: AC2 kV/min, 1mA

⑪ Pulse string interference (EFT): 2kV, 5kHz

⑫ Surge voltage: 2kV, 1.2/50μs

⑬ Insulation resistance: ≥100MΩ

The key operation instructions are shown in Table 3.2.4.

Section Ⅲ   Practical Training Project

Table 3.2.4   Key operation instructions

Display character	Corresponding definition	Data range
PASS	Accuracy adjustment	0~9999
bAUD	Baud rate	1200,2400,4800,9600,19200
dISP	Display value	0~19999
ADDr	Local address	1~32
rHNu	Relay output upper limit threshold	Outside range alarm when upper limit threshold＞ lower limit threshold
rLNu	Relay output lower limit threshold	Interval alarm when upper limit value ＜lower limit value Turn off alarm when upper limit threshold＝lower limit threshold
SAVE	Save and exit	
E	Do not save and directly exit	
r-rE	Return difference	
DECi	Decimal point position	

The menu will be increased or decreased according to the selected function. The operable keys are "SET", "←" and "→". Press the "SET" button to enter the debugger main menu, LED displays PASS (accuracy adjustment), press the "←" button to display dISP (display value), DECi (unit and decimal point position), SAVE (save and exit), E (do not save and directly exit), press the "→" button to enter reverse loop display, press the "←" or "→" button to reach the required setting item, press the "SET" button to enter the fixed item menu to be modified. When each of the following is adjusted to the desired value, press the "SET" button to return to the main menu. The detailed descriptions are as follows:

① Display value (dISP): LED displays 00000 (if the display position of the decimal point has been adjusted, the corresponding position shows the decimal point). At the same time, the modified place flashes, press "→" button to add 1 to the modified place, press the "←" button to move the modified place one place to the left, cycle like this and adjust to the required value (display value corresponds to the display value of the full input, for example: when DC100V is input, it can be displayed as 100.00).

② Decimal position (DECi): LED shows the current decimal point position, press the "→" button to move the decimal point one place to the right, move it to the last unit to indicate turning over (if there is no unit display, ignore). Press the "←" button, the decimal point moves one place to the left, and moves to the highest place unit to indicate turning over (if there is no unit display, ignore).

③ Local address (ADDr): LED flashing displays local address, press the "←" button to add 1 to the value, press the "→" button to minus 1 and adjust it to the corresponding value (1~32).

④ Baud rate (bAUD): LED flashing displays currently set baud rate (1200~19200), press the "←" button or "→" button to change the value.

⑤ The relay action threshold (rHNu/rLNu): LED displays the current action threshold while the modified place flashes, press the "→" button to add one to the modified place, press the "←" button to move the modified place one place to the left, cycle like this to adjust the value to the actual action threshold.

⑥ Relay action return difference setting (r-rE): LED displays the current action re-

25

turn difference (the return difference is the last two values of the actual measurement), at the same time, the modified place flashes, press the "→" button to add 1 to the modified place, press the "←" button to move the modified place one place to the left, cycle like this to adjust to the actual action return difference.

⑦ Accuracy adjustment (PASS): display accuracy adjustment, when modifying the accuracy, the corresponding password is needed, the factory has adjusted the accuracy, so users do not need to adjust it.

Press the "SET" button to save the modified items and to exit when the S is displayed. When displaying E, press the "SET" button to ignore the modification and to exit directly.

Description: under any level menu, after pressing the button, the time without operation is greater than 60 seconds, the system will automatically exit the setting menu.

The actual object picture is shown in Figure 3.2.6. The port definitions are shown in Table 3.2.5 and Table 3.2.6.

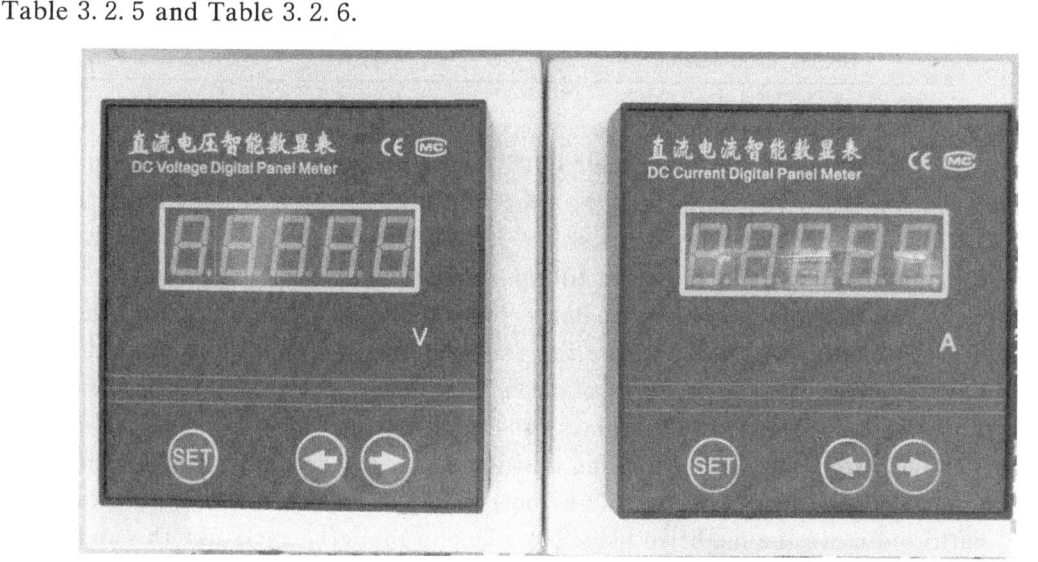

(a) Photovoltaic output voltage   (b) Photovoltaic output current

Figure 3.2.6   DC unit module

Table 3.2.5   DC voltmeter

S/N	Definition	Explanatory note
1	U+	Measured voltage input
2	U−	
3	A	RS-485
4	B	
5	L	AC 220V power input
6	N	

Table 3.2.6   DC ammeter

S/N	Definition	Explanatory note
1	I+	Measured current input
2	I−	
3	A	RS-485
4	B	
5	L	AC 220V power input
6	N	

(4) On-off sequence of circuit breaker

Note: the training is to apply adjustable resistance (Figure 3.2.7) as a load. Please refer to Table 3.2.7 to complete the training task.

Section Ⅲ Practical Training Project

Figure 3. 2. 7 Adjustable resistance

Table 3. 2. 7 On-off sequence of circuit breaker

S/N	Function	Powering-on sequence of circuit breaker	Powering-off sequence of circuit breaker	Powering-on sequence of adjustable resistance	Powering-off sequence of adjustable resistance
1	PV output	1	3	1	2
2	Adjustable load	×	×	2	1
3	PV MPPT	2	2	×	×
4	Storage battery	3	1	×	×

A rotary adjustable rheostat is used in the training. It can put the light source and photovoltaic panel in a fixed position. Then according to the resistance value from large to small, it can measure several groups of photovoltaic output currents and photovoltaic output voltages. After measuring a group of data, the angle of the light source can be changed to go on another test. Consider and compare the different characteristics of the volt-ampere characteristic data curve when the light source is at different angles.

【Project Principle and Basic Knowledge】

The energy converters of the solar photovoltaic power generation are solar batteries, which also known as photovoltaic batteries. When photovoltaic batteries are working, their terminal voltages will change with the variation of light intensity, ambient temperature and load, which makes the output power change greatly. The output of a photovoltaic battery has nonlinear characteristics and it is an unstable power supply. The training is to determine the volt-ampere characteristics of photovoltaic batteries in the current environment and find out the maximum output power point of photovoltaic batteries in the current environment under a fixed light intensity.

【Project Implementation】

① Close the "main power supply" switch on the "simulated energy control system" and then the system is powered on and the three-phase power indicator light will be on.

② Close the air switch of the "simulated light source" to make the simulated sun lamp turn on.

③ Close the "main power supply" switch on the "energy conversion

Solar panel *U-I* characteristic test

27

storage control system" and then the system is powered on and the three-phase power indicator light will be on.

④ Record the PV output voltage and this value is the PV open-circuit voltage.

⑤ Close the air switches of the "photovoltaic output" and the "adjustable load" on the "energy conversion storage control system".

⑥ Adjust the adjustable resistance to measure the several groups of photovoltaic output currents and photovoltaic output voltages according to the resistance value from large to small and record them in Table 3. 2. 8.

Table 3. 2. 8   Volt-ampere characteristics of photovoltaic batteries

S/N	Voltage/V	Current/A	Power/W	Remark
1	0			Short-circuit current
2	1			
3	2			
4	3			
5	4			
6	5			
7	6			
8	7			
9	8			
10	9			
11	10			
12	11			
13	12			
14	13			
15	14			
16	15			
17	16			
18	17			
19	18			
20	19			Open-circuit voltage

⑦ By recording the voltage and current data, calculate the corresponding power of each voltage and current.

⑧ After the experiment is finished, turn off the air switch of the "simulated light source" and the main power switch on the simulated energy control system, the air switches of the "photovoltaic output" and the "adjustable load" and the main power switch on the energy conversion storage control system in turn. If a subsequent experiment will be carried out, the main power switch can not be turned off.

【Project Exercise】

① Draw the U-I curve and the power curve according to the measured data, and find our the maximum power point.

② Calculate which point the maximum power output point is and why?

28

Section Ⅲ Practical Training Project

Project Ⅲ Design of System Components and Battery Capacity

【Project Description】

Learn the key contents of solar photovoltaic power generation technology and grasp the principles and basic knowledge of solar cell components and storage battery selection. Be able to calculate the capacity and quantity of the solar cell component according to actual environmental conditions and can rationally design the capacity and quantity of the storage battery under different conditions.

【Ability goals】

① To calculate the capacity and quantity of solar cell components as required.
② To calculate the capacity and quantity of the storage battery as required.

【Project Environment】

The design principle of solar cell component is to meet the demand for bearing daily power consumption under average weather conditions (solar radiation), that is to say, the full-year power generation of solar cell component is equal to the amount bearing annual power consumption. Due to the fact that weather conditions may be lower or higher than the average, solar cell components are designed to meet the needs of the worst-illumination and least-solar-radiation seasons. If it is designed only on an average basis, it will result in a continuous loss of battery power in the worst-illumination season for more than a third of the year. Long-term depletion of the battery will cause the sulfation of polar plates, making the service life and performance be greatly affected, and the follow-up operating costs of the system will be greatly increased. It is also not advisable to design a solar cell component too large for the purpose of filling the battery as quickly as possible, so that it may generate far more power than it should bear during the most of the year, resulting in waste of solar cell components and higher overall system costs. Therefore, the best way to design solar cell components is to make them basically meet the needs of the worst-illumination seasons, namely, the battery can also be basically full of electricity every day in the worst-illumination seasons.

In some areas, the season with worst-illumination is far below the full-year average, and if the power of solar cell components is still designed based on the worst situation, the amount of electricity generated at other times of the year will far exceed what is actually needed, creating a waste. At this point, what can only be done is to properly increase the design capacity of the battery, increase the storage of electrical energy, and make the battery in a shallow discharge state, making up for the damage caused by the insufficient power generation in the worst-illumination season. Place where is allowed by actual condition still can consider to take measures such as complementarity between wind power generation and solar power generation (which is referred to as wind-solar complementary power generation) and mains electricity complementarity to achieve the best overall comprehensive cost-effectiveness of the system.

29

(1) Design methods for solar cell components and arrays

The basic method to design and calculate the size of a solar cell component is to use the data loading the average daily energy consumption (unit: ampere-hour or watt-hour) as the basic data, the local solar radiation resource parameters such as peak sunshine hours and total amount of annual radiation as reference, so as to calculate synthetically by combining the data of some related factors or coefficient.

There are generally two ways to design and calculate a solar cell component or a component array. One method is to calculate the power of the the solar cell component or the array directly based on the above data, and to select or customize the corresponding cell component according to the calculated results, thus obtaining the shape size and installation size of the cell component. This method is generally applicable to the design of small and medium-sized photovoltaic power generation systems. The other method is to select a solar cell component which meets the requirements in size, based on the peak power data, the peak working current and the daily power generation, then to carry out calculation by combining the above data, and finally to determine the series, parallel number and total power of the cell component in the calculation. This method is suitable for the design of medium-scale and large-scale photovoltaic power generation systems. The second method is taken as an example to introduce the design formula and method of a common solar cell component.

(2) Basic calculation method for solar cell components and arrays

The basic way to calculate the solar cell component is that the average daily load of power consumption (Ah) is divided by the average daily power generation (Ah) of the selected cell component in a day, thus calculating the number of solar cell components that need to be connected in parallel throughout the system. The parallel output current of these solar cell components is the current required for the system load. The specific formula is as follows:

Number of parallel connection of solar cell components

$$= \frac{\text{average daily load of electricity consumption (Ah)}}{\text{average daily power generation of cell components (Ah)}} \qquad (3.3.1)$$

average daily power generation of cell components

$$= \text{peak working current (A)} \times \text{peak sunshine hours (h)}$$

The number of solar cell components in series can be calculated by dividing the operating voltage of the system by the peak operating voltage of the solar cell component. When these cell component are in series connection, the operating voltage required for the system load or the charging voltage for the battery pack can be generated. The specific formula is as follows:

Number of series connection of solar cell components

$$= \frac{\text{operating voltage of the system (V)} \times 1.43}{\text{peak operating voltage of the cell component (V)}} \qquad (3.3.2)$$

The coefficient 1.43 is the ratio of the peak operating voltage of a solar cell component to the operating voltage of the system. For example, the peak voltage of a solar cell component that supplies or charges a system with a working voltage of 12V is $17\sim17.5$V (12×1.43); while the peak voltage that supplies or charges a system with a working voltage of 24V is $34\sim34.5$V (24×1.43). Therefore, for the convenience of calculation, the working voltage of the system is multiplied by 1.43 is the approximate value of the peak voltage of the component or the entire array. For example, assuming that a PV power generation system has a working voltage of 48V, a cell component with a peak operating vol-

tage of 17.0V is selected, the series connection number of the cell components is calculated, that is, $48V \times 1.43/17.0V = 4.03 \approx 4$.

With the parallel connection number and the series connection number of the cell components, the total power of the cell components or the array can be easily calculated. The calculation formula is as follows:

Total power(W)of the cell component(array)
= the number of parallel connection of the components $\times$
the number of series connection of the components $\times$
the peak output power(W)of the selected component　　　(3.3.3)

(3) Design method for storage battery and battery pack

The storage battery aims at ensuring the normal power load of the system when the solar radiation is insufficient. To ensure the normal operation of the system in a few days, it is necessary to introduce a meteorological condition parameter: namely, continuous rainy days. Generally, the design parameter is based on the local maximum continuous rainy days, but the requirements of the load on the power supply should also be taken into consideration. In view of general load such as solar street lamp, the design parameter can be selected within 3~7 days according to experience or need. While for important loads such as communications, navigation, hospital care, etc., the design parameter is selected within 7~15 days. In addition, the installation location of a photovoltaic power generation system should also be considered. The capacity of the batteries should be designed larger for a remote place, because it will take much time for the maintenance personnel to arrive. In practical application, some mobile communication base stations are extremely inconvenient to go because of the high mountains and long roads, so they are equipped with a set of spare battery pack in case of emergency while arranging the normal battery pack. This kind of generation system not considers economical benefit simply, but gives priority to reliability.

The design of the storage battery mainly includes the design calculation of the capacity of the storage battery and the design of the series-parallel connection combination. In photovoltaic power generation system, lead-acid battery is used widely, which mainly takes into account the mature technology and cost, so the design and calculation method introduced below is mainly based on lead-acid battery.

(4) Basic calculation method of storage battery and battery pack

The initial battery capacity can be obtained by multiplying the amount of electricity load needed per day by continuous rainy days based on local meteorological data or actual conditions. The capacity of the battery is then divided by the maximum discharge depth coefficient allowed by the battery. Because of the characteristics of the lead-acid battery, it can not be used up by 100% discharge during definite continuous rainy days, otherwise the battery will be scrapped in a short time and be greatly shorten its service life. It is necessary to be divided by the maximum discharge depth coefficient to obtain the required battery capacity. The selection of the maximum discharge depth requires to refer to the performance parameter data provided by the battery manufacturer. Generally speaking, 50% discharge depth is chosen for shallow cycle battery, while 75% discharge depth is chosen for deep cycle battery. The basic formula for calculating the capacity of a battery is as follows:

Capacity of the stroage battery
$$= \frac{\text{average daily load of electricity consumption (Ah)} \times \text{continuous rainy days}}{\text{maximum discharge depth}} \quad (3.3.4)$$

【Project Principle and Basic Knowledge】

(1) Consideration of related factors in the calculation of solar cell component and array

The basic calculation method of a solar cell component and a array is completely written calculation under ideal condition. The capacity of the cell component calculated according to the above calculation formula can't meet the power demand of the photovoltaic power generation system in practical application. To get more accurate data, some relevant factors and data should be taken into account and included in the calculation.

There are two main factors related to the power generation of the solar cell component.

① Power decay of the solar cell component. In the practical application of photovoltaic power generation system, the output power (power generation capacity) of the solar cell component will be attenuated or decreased by various internal and external factors, such as the coverage of dust, the power decay of the component itself, the wear and tear of the line and other non-quantifiable factors. Moreover, the conversion efficiency of the AC inverter should be considered in the AC system. Consequently, in design, the factors causing the power decay of the cell component are calculated by 10% loss, and the loss of the conversion efficiency of AC inverter is also calculated by 10% loss for AC photovoltaic power generation system. These are actually safety factors that need to be taken into account in the design of a photovoltaic power generation system, leaving a reasonable margin for the cell componet when designing, which is the guarantee for the long-term normal operation of the system.

② The charge and discharge loss of the battery. During the charging and discharging process of the battery, the current generated by the solar cell will be lost due to heat, evaporation of electrolyzed water and so on in the process of transformation and storage, that is to say, the charging efficiency of the battery is usually only 90%~95% depending on the different batteries. Therefore, the power of the battery pack should be increased by 5%~10% according to the difference of the battery when designing to offset the dissipation loss during the charging and discharging of the battery.

(2) Practical calculation formula of solar cell component and array

After taking all factors into account, the correlation coefficient is incorporated into the formulas (3.3.1) and formulas (3.3.2), which is a complete formula for designing and calculating a solar cell component.

Dividing the average daily load of electricity consumption by the charging efficiency of the battery will increase the daily load of electricity consumption, which actually gives the actual load that the cell component needs to load. By multiplying the loss coefficient of the cell component by the average daily power generation capacity of the cell component, the reduction of component power generation caused by environmental factors and component self-degradation is taken into account, thus providing a conservative estimate of the amount of the power generation by a solar cell used in practice. Taking into account the above factors, the calculation formula is as follows:

Parallel connection number of cell components

$$= \frac{\text{average daily load of electricity consumption (Ah)}}{\text{average daily power generation of the component (Ah)} \times \text{charging efficiency coefficient} \times \text{the loss coefficient of the component} \times \text{inverter effciency coefficient}}$$

(3.3.5)

Series connection number of cell components

$$= \frac{\text{operating voltage of system} \times 1.43(\text{coefficient})}{\text{peak operating voltage of the component}} \qquad (3.3.6)$$

Total power(W)of cell component(array)

$$= \text{parallel connection number of the components} \times$$
$$\text{series connection number of the components} \times$$
$$\text{the peak output power(W)of the selected components} \qquad (3.3.7)$$

When designing and calculating the solar cell component, the effect of seasonal variation on power generation of the system should be considered. The reason is that the design and calculation of the capacity of components is generally based on the local solar radiation resources parameters such as peak sunshine hours, annual total radiation data, which are the annual average data, there is no problem for spring, summer and fall by referring to the calculation result based on above data, while it may be not appropriate for winter. Therefore, it is advisable to calculate monthly power generation based on the solar radiation resource parameters per month in the local area when conditions are available or when designing a more important photovoltaic power generation system, the maximum of which is the number of cell components needed during the year. For example, it is calculated that the number of solar cell components needed in the winter is eight somewhere, but maybe five is enough in the summer. It is better to choose the number of solar cell components in the winter to keep the system running properly throughout the year.

(3) Considerations on the related factors of the basic calculation method of storage battery and battery pack

The basic formula of storage battery and battery pack is only the basic method to estimate the capacity of storage battery, while in practical application, some performance parameters will influence the capacity and service life of storage battery, mainly including the discharge rate of the battery and the operating environment temperature.

① Effect of discharge rate on capacity of storage battery. Firstly, the concept of battery discharge rate is briefly reviewed. The so-called discharge rate is the ratio of discharge time and discharge current to battery capacity, which is divided into 20-hour rate (20h), 10-hour rate (10h), 5-hour rate (5h), 3-hour rate (3h), 1-hour rate (1h), 0.5-hour rate (0.5h) and so on. When it discharges with large current, the discharge time is short, and the capacity of the battery will be lower than the nominal capacity. When it discharges with small current, the discharge time is long, and the actual discharge capacity will be larger than the nominal capacity. For example, a battery with a capacity of 100Ah can use a current discharge of 2A for 50 hours, while a current discharge of 50A may take no more than 2 hours, so the actual capacity is not 100Ah enough. The capacity of the battery varies with the discharge rate, which can affect the capacity design. When the load discharge current of the system is large, the actual capacity of the storage battery will be smaller than that of the design capacity, which will lead to the shortage of power supply. The actual capacity of the battery will be larger than the designed capacity when system loads small operating current, which will lead to a pointless increase of the system cost. In particular, the battery used in a photovoltaic power generation system, its discharge rate is generally slow, almost in the 50-hour rate or above, while the battery nominal capacity provided by manufacturer is below 10-hour discharge rate capacity. Hence, the influence factors of battery discharge rate on capacity in a photovoltaic power generation system

should be taken into account in the design, and the actual average discharge rate of photovoltaic system should be calculated. According to the capacity of the battery provided by the manufacturer at different discharge rates, the capacity of the battery can be calibrated and corrected. When there is no detailed capacity-discharge rate data at hand, it is also available to estimate the capacity of the battery with the slow discharge rate $50 \sim 200h$ (hour rate) in the photovoltaic system. The capacity of the battery can be increased by 5% $\sim 20\%$ compared with the standard capacity of the battery, and the correction coefficient of the corresponding discharge rate is $0.95 \sim 0.8$. The formula for calculating the average discharge rate of the photovoltaic system is as follows:

$$\text{Average discharge rate} = \frac{\text{number of continuous rainy days} \times \text{working time of load}}{\text{maximun discharge depth}}$$

(3.3.8)

For PV systems with many different loads, the load working time needs to be calculated by the weighted average method. The calculation method of the weighted average load working time is as follows:

$$\text{Working time of load} = \frac{\Sigma \text{power of load} \times \text{working time of load}}{\Sigma \text{power of load}}$$

(3.3.9)

According to the above two formulas, the actual average discharge rate of the photovoltaic system can be calculated, and the capacity of the battery can be modified according to the capacity of the battery provided by the manufacturer at different discharge rates.

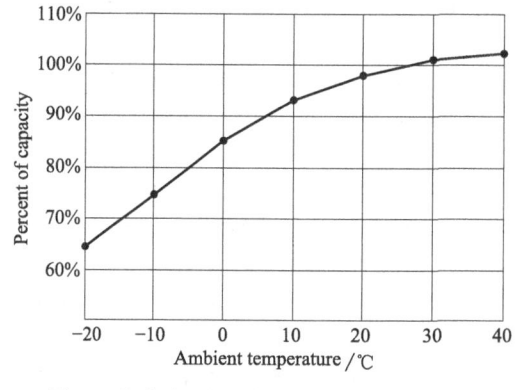

Figure 3.3.1 Graph of relation between battery temperature and discharge capacity

② Effect of ambient temperature on capacity of storage battery. The capacity of the storage battery will change with the temperature of the storage battery. When the temperature of the storage battery drops, the capacity of the storage battery will decrease, and when the temperature is below zero, the capacity of the storage battery will decrease sharply. When the temperature rises, the capacity of the batter will slightly increase accordingly. The relationship between battery temperature and discharge capacity is shown in Figure 3.3.1. The nominal capacity of the battery is generally calibrated at the ambient temperature of 25℃. With the decrease of temperature, the capacity of the battery decreases to $95\% \sim 90\%$ of the nominal capacity at 0℃, to $90\% \sim 80\%$ of the nominal capacity at -10℃, and to $80\% \sim 70\%$ of the nominal capacity at -20℃, so the influence of the ambient temperature of the battery on its capacity must be taken into account. When the minimum temperature is too low, some measures should be taken to keep the battery warm, such as burying in the ground, moving into the room, or using colloid lead-acid batteries with higher price.

When the minimum temperature in the installation location of the PV system is very low, the capacity of the batteries needed in the design is larger than that of the normal temperature range, so that the PV system can provide the required energy at the lowest temperature. Therefore, when designing, the correction coefficient can be found from the temperature-capacity correction diagram of the battery provided by the manufacturer,

Section Ⅲ   Practical Training Project

which can be used to calculate the capacity of the battery by the formula. If there is no corresponding temperature-capacity correction curve, the temperature correction coefficient can be determined according to experience. Generally, the correction coefficient can be selected among 0. 95~0. 9 when it is 0℃, among 0. 9~0. 8 when it is -10℃, and among 0. 8~0. 7 when it is -20℃.

Furthermore, low ambient temperature can also affect the maximum discharge depth. When the ambient temperature is below -10℃, the maximum discharge depth of the shallow cycle battery can be adjusted from 50% at room temperature to 35%~40%, and the maximum discharge depth of the deep cycle battery can be adjusted from 75% at room temperature to 60%. In this way, while improving the life of the battery and reducing the maintenance cost of the battery system, the system cost will not be too high.

(4) Practical calculation formula of battery capacity

After taking all factors into account, the correlation coefficient is included in the above formula, which is a practical and complete formula to design and calculate the capacity of the battery. That is,

The capacity of the battery

$$= \frac{\text{average daily load of electricity consumption (Ah)} \times \text{continuous rainy days} \times \text{correction coefficient of discharge rate}}{\text{maixmum discharge depth} \times \text{correction coefficient at low tempatature}}$$

$$(3. 3. 10)$$

When the required capacity of the battery is determined, the series-parallel connection design of the battery pack should be carried out. The calculation method of the series-parallel combination of the battery pack is introduced below. Batteries are provided with nominal voltage and nominal capacity, such as 2V, 6V, 12V and 50Ah, 300Ah, 1200Ah, etc. In order to achieve the operating voltage of the system, it is necessary to connect the battery in series to supply the system and the load. The number of batteries in series is to divide the working voltage of the system by the nominal voltage of the selected battery. The number of batteries that need to be parallel is to divide the total capacity of the battery pack by the nominal capacity of the selected battery monomer. The nominal capacity of the battery monomer can have a variety of options. For example, if the calculated capacity of the battery is 600Ah, then we can choose one 600Ah single battery, or we can choose two 300Ah batteries in parallel connection, or choose three 200Ah or six 100Ah batteries in parallel connection. In theory, none of these options are problematic. However, in practice, it is recommended to choose large-capacity batteries to reduce the number of parallel batteries. The purpose is to minimize the impact of the imbalance between the batteries. The more parallel groups there are, the greater the possibility of battery imbalance. Generally speaking, the number of parallel batteries should not exceed 4. The calculation formula of the number of batteries in series and parallel connection is as follows:

$$\text{Number of batteries in series connection} = \frac{\text{operating voltage of the system}}{\text{nominal voltage of the battery}}$$

$$(3. 3. 11)$$

$$\text{Number of batteries in parallel connection} = \frac{\text{total capacity of batteries}}{\text{nominal capacity of the battery}}$$

$$(3. 3. 12)$$

Selection of solar
panel and battery

【Project Implementation】

① Calculation   A solar photovoltaic power supply system for a mobile communication base station is built somewhere. The system uses DC load with load operating voltage of 48V, and the daily power consumption of 150Ah. The lowest solar radiation in the area is in January, and its peak sunshine hours are 3.5h in the inclined plane. 125W solar cell component is selected, its main parameters are: peak power 125W, peak working voltage 34.2V, peak working current 3.65A, please calculate the number of solar cell components used and the combination design of solar cell array.

② Calculation   A solar photovoltaic power supply system for a mobile communication base station is built somewhere, which adopts DC load with load operating voltage of 48V. The system has two sets of equipment load: the working current of one set of equipment is 1.5A, operating 24h a day; the working current of the other set of equipment is 4.5A, operating 12h a day. The minimum temperature in this area is -20℃, and the maximum continuous rainy days are 6. Deep cycle batteries are used. Please calculate the capacity of the battery pack, the number of batteries in series connection and in parallel connection, and the connection method.

【Project Exercise】

① According to the parameters and requirements listed in the experimental steps, calculate the number of solar cell components used, the combination mode of solar cell arrays and series-parallel connection schematic diagram.

② According to the parameters and requirements listed in the experimental steps, calculate the capacity of the battery pack, the number of series-parallel connections and the schematic diagram of connection mode.

③ What are the meanings of calculating the capacity and quantity of storage batteries?

Project Ⅳ   Wind Turbine Characteristics Curve Test

【Project Description】

Learn the working characteristics of wind turbines; be familiar with the factors affecting the output power of wind turbines and lay the foundation for theoretical analysis to improve the power generation efficiency of wind turbines.

【Ability goals】

① To learn the working characteristics of wind turbines.

② To be familiar with the factors that influence the output power of wind turbines and to lay a good foundation for theoretical analysis to improve the efficiency of wind turbines.

【Project Environment】

To complete this training task, we should refer to the training platform equipment manual of THWPFG-4 wind-solar complementary power generation system, learn the electrical schematic diagram of the simulated

Introduction of wind
turbine generator

Section Ⅲ Practical Training Project

wind energy control unit.

The circuit diagram of the simulated wind energy control unit is shown in Figure 3. 4. 1.

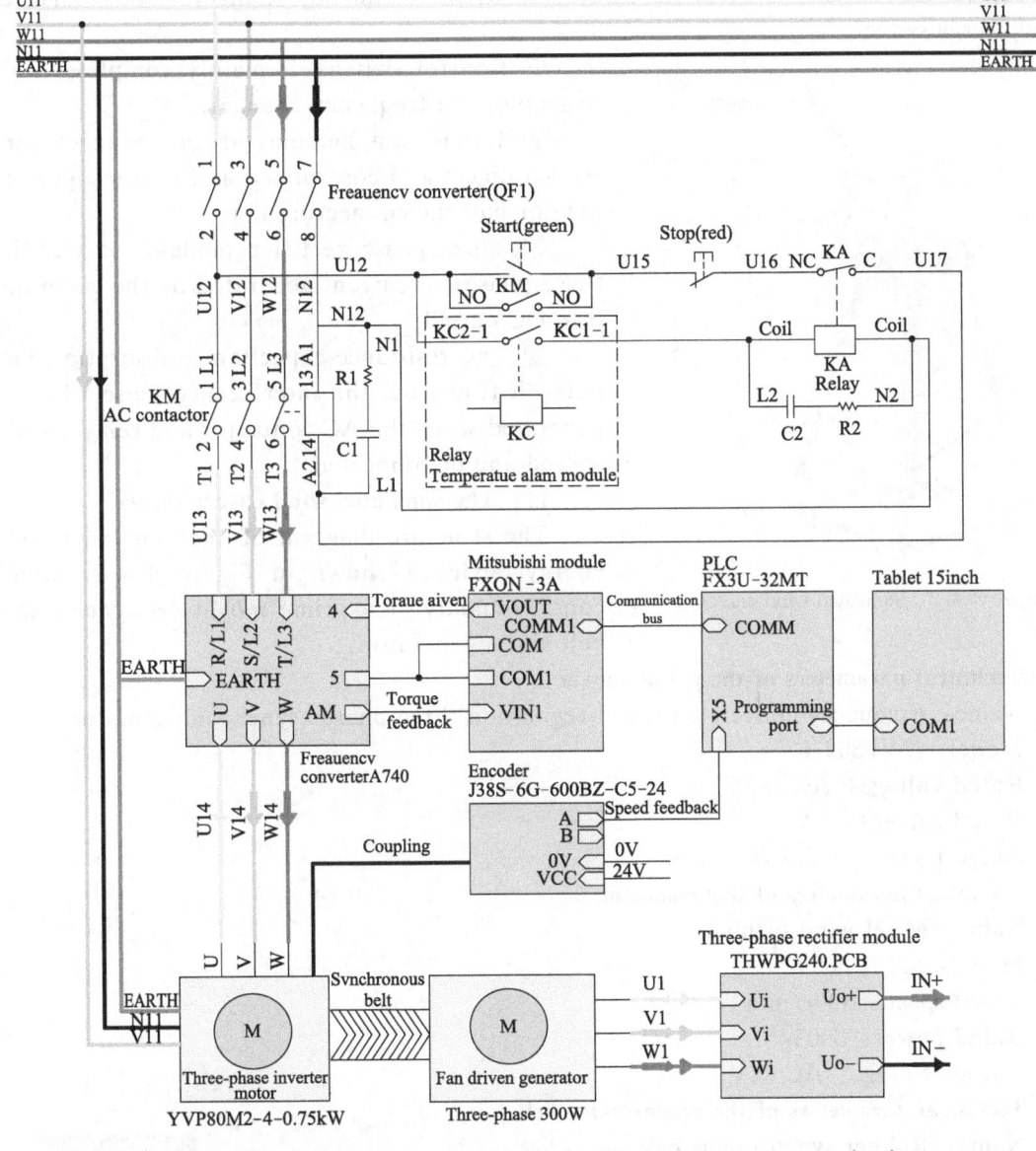

Figure 3. 4. 1 Circuit diagram of the simulated wind energy control unit

The simulated wind energy control unit consists of stimulated wind energy device, frequency changer, AC contactor, relay, control switch, connection module, three-phase rectifier module and resistance-capacitance absorption module.

① Stimulated wind energy device: the simulated wind energy device is mainly composed of a prime mover and a generator, and the prime mover drives the generator to rotate through the synchronous belt to simulate the process of wind power generation.

② Frequency changer: it, by changing the frequency and amplitude of AC motor power supply, achieves smooth control of motor speed.

③ AC contactor: in the stimulated wind energy system, it, by combining with electromagnetic relay and control switch, controls start and stop of the frequency changer.

37

④ Electromagnetic relay: it can carry out the isolation of signals, automatic regulation, safety protection, circuit conversion and so on, and by cooperating with AC contactor and control switch, control start and stop of the frequency changer in the stimulated wind energy system.

⑤ Control switch: it mainly completes start or stop of the frequency changer.

⑥ Connection module: it can connect controlled object and controller, and is convenient to plug or pull the connecting line.

⑦ Three-phase rectifier module: it rectifies three-phase AC current generated by the generator into DC output.

⑧ The resistance-capacitance absorption module: it filters out the interference pulse which is generated when the AC contactor and relay are absorbed and disconnected.

(1) The simulated wind energy device

The structure diagram of the simulated wind energy device is shown in Figure 3.4.2, which consists mainly of a prime mover, a synchronous belt and a generator.

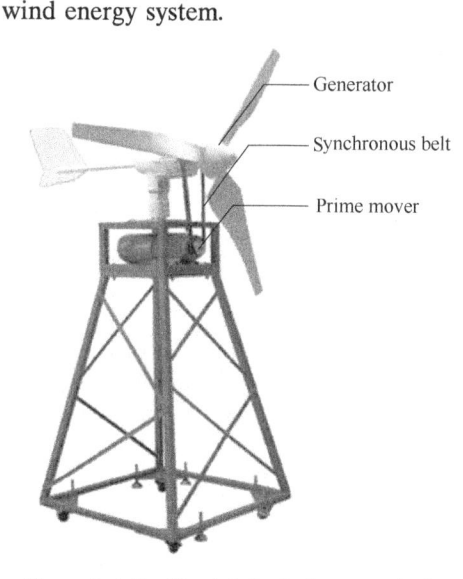

Figure 3.4.2　Simulated wind energy device

Technical parameters of the prime mover

Name: Frequency-conversion speed-regulation three-phase asynchronous motor

Model: YVP802-4

Rated voltage: AC 380V

Rated current: 2A

Rated power: 0.75kW

Technical parameters of the generator

Name: MLH wind turbine

Model: MLH-300W

Rated speed: 400r/min

Rated power: 300W

Output voltage: AC 24V

Technical parameters of the synchronous belt

Name: Rubber synchronous belt

Model: 424XL

Junction line length: 1076.90mm

Tooth quantity: 212

(2) The frequency changer (Figure 3.4.3)

The frequency changer can optimize the operation of the motor, so it can also improve the efficiency and save energy.

Technical parameters

Power range: 0.4~0.75kW

General vector control, when it reaches 1Hz, 150% torque output is generated

Long life components are adopted

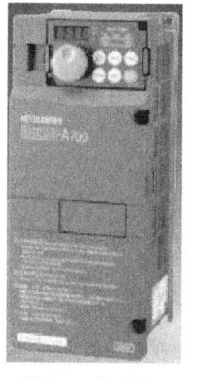

Figure 3.4.3
Frequency changer

Section Ⅲ　Practical Training Project

Built-in modbus -RTU protocol
Built-in brake transistor
Extended PID，triangular wave function
Having safety stop function
The Structure of the Frequency changer（Figure 3. 4. 4）

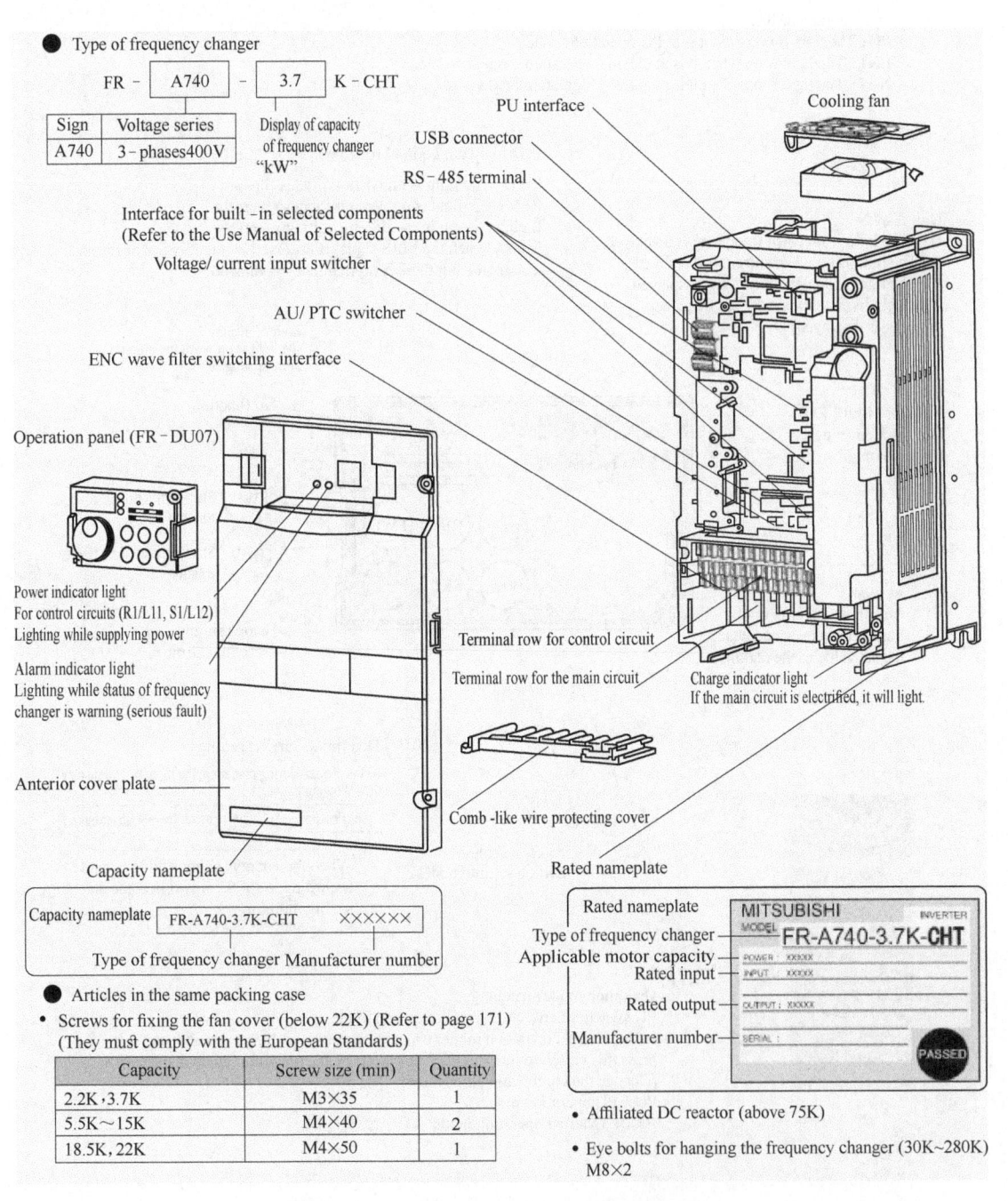

Figure 3. 4. 4　Structure of frequency converter

Parameter setting of the frequency changer

39

① Name of each part of the operation panel（Figure 3. 4. 5）
② Basic operation（set value when leaving the factory，Figure 3. 4. 6）

Operation mode dispaly
PU：The light is on when it is at PU operation mode
EXT：The light is on when it is at external operation mode
NET：The light is on when it is at network operation mode

Display direction of rotation
FWD : The light is on when it rotates forward
REV : The light is on when it rotates reversely
Light is on：it is rotating forward or reversely
Blink : When the MRS signal is accessed by the forward or reverse
command, but there is no frequency command.

Unit display
Hz：The light is on when frequency is displayed
A : The light is on when current is displayed
V : The light is on when voltage is displayed
(The light twinkles when the set frequency
monitor is displayed)

Monitor display
The light is on when the monitor
mode is adopted

Monitor (4-bit LED)
Display frequency,
reference number, etc.

No function

M knob
Mitsubishi inverter knob
Set the frequency and change
the set value of the parameter

FWD   Start instruction of reverse
rotation

REV   Start instruction of reverse
rotation

STOP
RESET   Stop running
reset to alarm

SET   Determine various settings
If it is pressed during operation, the monitor will display
the following in cycles

| Operating frequency | → | Output current | → | Output voltage | * |

MODE
Mode switch
Switch setting modes

* When the energy saving setting is made
according to Pr. 52, it will become an energy
saving monitor.

PU
EXT

Operation mode switching
PU switches between external operation mode
In the external operation mode (using a separately set frequency and start signal),
press this button to turn the EXT displayed by the operation mode display on.
(Change the combination mode to Pr. 79)
PU : PU operation mode
EXT : External operation mode

Figure 3. 4. 5   Operation panel of the frequency changer

40

Section Ⅲ   Practical Training Project

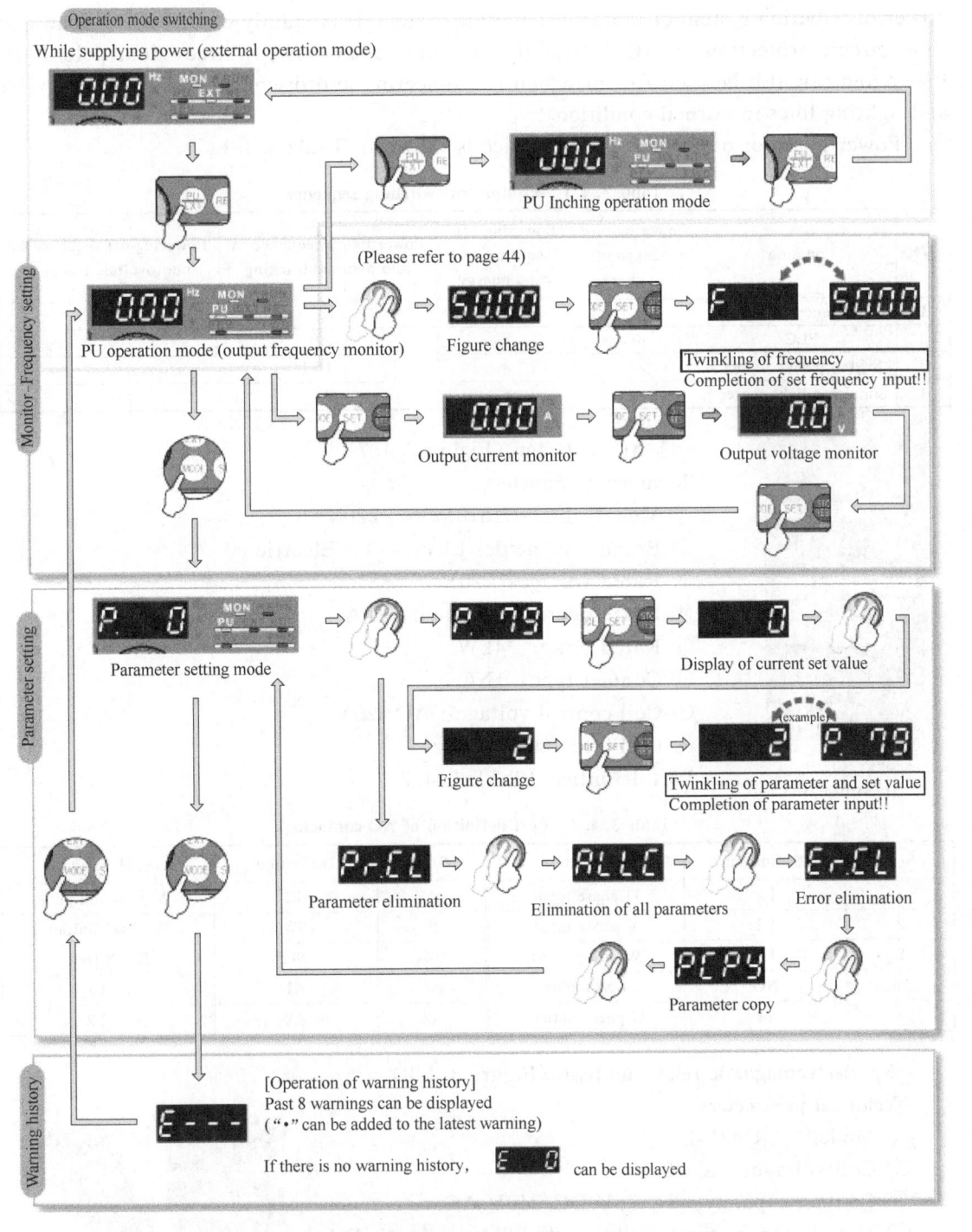

Figure 3.4.6   Basic operation of frequency changer

(3) Circuit breaker

It refers to a mechanical switch that can connect and disconnect normal current in a

load-carrying circuit, and can connect and disconnect carrying current in a specified abnormal circuit condition (e. g. short circuit) and in a given time. It is suitable for lighting power distribution system or motor distribution system. It is mainly used for over-load and short-circuit protection of AC 50Hz/60Hz unipolar 230V, two, three, four-pole 400V lines, and can also be used for infrequently connecting and disconnecting electric devices and lighting lines in normal condition.

Powering on or off order of each device is shown in Table 3. 4. 1.

Table 3. 4. 1 Equipment switching sequence

S/N	Function	Powering-on sequence of wind energy	Powering-off sequence of wind energy	Powering-on sequence of photovoltaic tracking	Powering-off sequence of photovoltaic tracking
1	Frequency changer	1	1	×	×
2	PLC	×	×	2	2
3	Switch power supply	×	×	1	3
4	Simulated light source	×	×	3	1

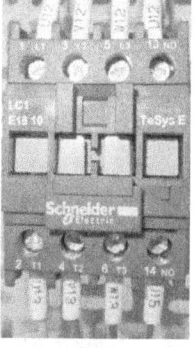

Figure 3. 4. 7
AC contactor

(4) AC contactor (Figure 3. 4. 7)

Technical parameters

① Model: LC1-D1810M5N 220V

② Brand: Schneider Electric-TE Electric

③ Rated voltage: 220V

④ Rated operating current: 18A

⑤ Rated power: 4kW

⑥ Contact type: 1NC

⑦ Coil control voltage: AC 220V

⑧ Coil frequency: 50Hz

Port definition (Table 3. 4. 2)

Table 3. 4. 2 Port definition of AC contactor

S/N	Definition	Explanation	S/N	Definition	Explanation
1	L1	U phase input	4	T2	V phase output
3	L2	V phase input	6	T3	W phase output
5	L3	W phase input	14	NO	KM 2 foot
13	NO	KM 1 foot	A1	A1	Coil 1
2	T1	U phase output	A2	A2	Coil 2

(5) electromagnetic relay and base (Figure 3. 4. 8)

Technical parameters

① Model: ARM2F-L

② Coil voltage: AC 220V

③ Contact capacity: 5A/28V DC/240V AC

④ Installation methods: Plug-and-pull type (with light for indication)

Port definition (Table 3. 4. 3)

Table 3. 4. 3 Port definition of electromagnetic relay and base

Figure 3. 4. 8 Electromagnetic relay and base

Section Ⅲ　Practical Training Project

S/N	Definition	Explanation	S/N		Definition
1	NC	Usually closed contact	9		C
4	NC	Usually closed contact	12		C
5	NO	Usually open contact	13		Coil
8	NO	Usually open contact	14		Coil

(6) control switch (Figure 3.4.9)

It controls start-up and stop of the frequency changer, with "green" button for start and "red" button for stop.

(7) connection module (Figure 3.4.10)

A connection module plays a connecting role between the controlled object and the controller for convenient plugging in or pulling out the linking wires, providing two 10-core connectors, and at most 20 connecting lines if necessary.

(8) three-phase rectifier module (Figure 3.4.11)

It can rectify the three-phase alternating current from the generator to the direct current output.

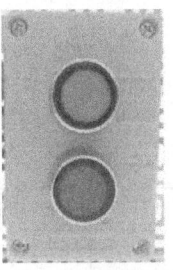

Figure 3.4.9
Control switch

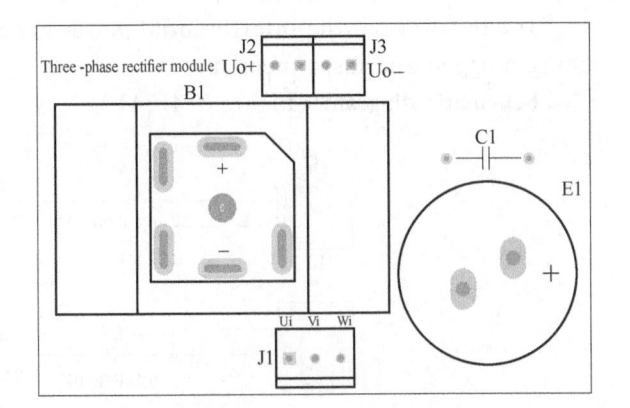

Figure 3.4.10　Connection module

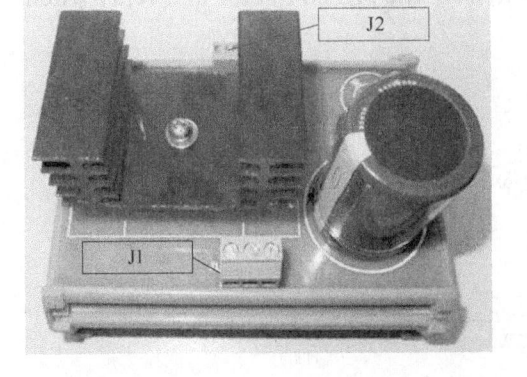

Figure 3.4.11　Three-phase rectifier module

Schematic diagram (Figure 3.4.12)

Port definition (Table 3.4.4)

43

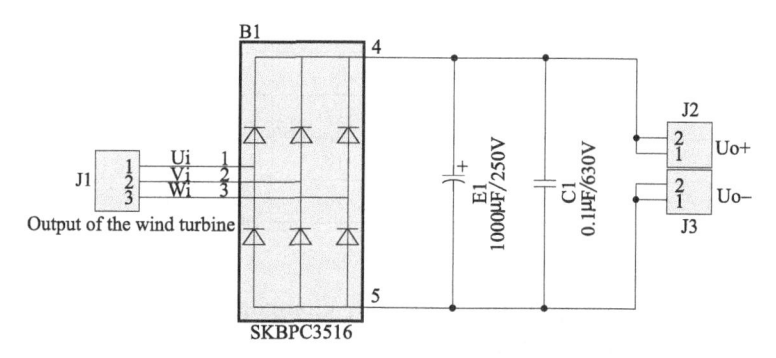

Figure 3.4.12   Schematic diagram of three-phase rectifier module

Table 3.4.4   Port definition of three-phase rectifier module

S/N	Definition	Explanation	S/N	Definition	Explanation
1	J1: Ui	Generator U-phase output	3	J1: Wi	Generator W-phase output
2	J1: Vi	Generator V-phase output	4	J2: Uo+	DC output

(9) resistance-capacitance absorption module (Figure 3.4.13)

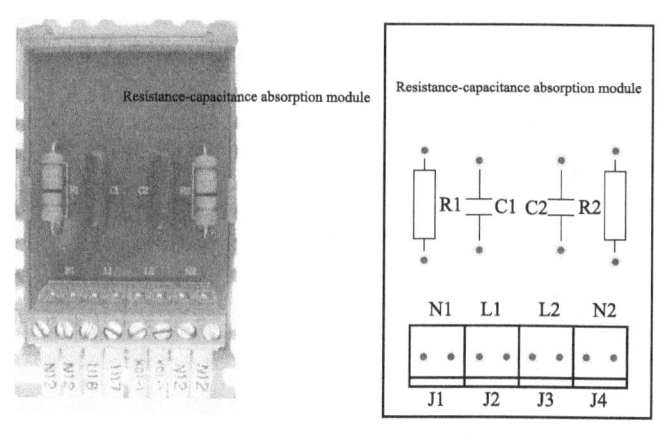

Figure 3.4.13   Resistance-capacitance absorption module

It can filter out the interferential pulses generated by the AC contactor and relay when being plugged and disconnected.

Schematic diagram (Figure 3.4.14)

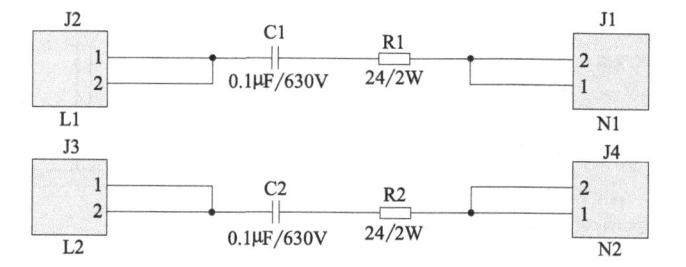

Figure 3.4.14   Schematic diagram of resistance-capacitance absorption module

Port definition (Table 3.4.5)

Table 3.4.5   Port definitions of resistance-capacitance absorption module

Section Ⅲ Practical Training Project

S/N	Definition	Explanation	S/N	Definition	Explanation
1	J1:N1	AC contactor coil 1	3	J3:L2	Solenoid relay coil 1
2	J2:L1	AC contactor coil 2	4	J4:N2	Solenoid relay coil 2

(10) rotary encoder
Model specification（Table 3.4.6）

Table 3.4.6 Model description of rotary encoder

J38S	-06	-1000	B	Z	-C	5-24
Shell size	Exit shaft	Resolution ratio	Output phase	Zero signal	Output mode	Operating voltage:
φ38mm	05 = φ5mm 06 = φ6mm 08 = φ8mm	10-2500P/R	B:2-phase output A,B	S = no Z signal M = "1" with Z signal output N = "0" with Z signal output	T = voltage type output NPN + R C = NPN open collector P = complementary circuit NPN + PNP L = long line driver (26 LS31) O = long line driver(7272) V = Long line driver OC (7273)	5 = + 5V 526 = + 5～ + 26V

Electrical parameters（Table 3.4.7）

Table 3.4.7 Electrical parameters of the rotary encoder

Product type	Incremental	Response frequency	300 kHz
Power supply voltage	5V DC, 5～26V DC	Output voltage	H:VCC×70%, L:0.8V
Consumption current	≤150mA		

Mechanical and environmental parameters（Table 3.4.8）

Table 3.4.8 Mechanical and environmental parameters of the rotary encoder

Maximum RPM	6000r/min	Applicable temperature	− 30～ + 85 ℃
Starting torque	0.05N·m	Storage temperature	− 35～ + 95 ℃
Radial load	50N	Resistance to impact	50g, 11ms
Axial load	20N	Quake-resistance	10g, 10～2000Hz

Wiring table（Table 3.4.9）

Table 3.4.9 Rotary encoder wiring table

Signal	A	B	Z	A-phase inversion	B phase inversion	Z-phase inversion	VCC	0V
Colour	Green	White	Yellow	Brown	Gray	Orange	Red	Black

(11) mechanical and environmental parameters
Mitsubishi module introduction

FXON-3A stimulated special function module has two input channels and one output channel. The input channel receives the stimulated signal and converts the stimulated signal

into digital value. The output channel adopts digital value and outputs equivalent stimulated signal. The maximum resolution of FXON-3A is 8 bits.

Terminal layout and wiring

Terminal layout and wiring are shown in Figure 3. 4. 15.

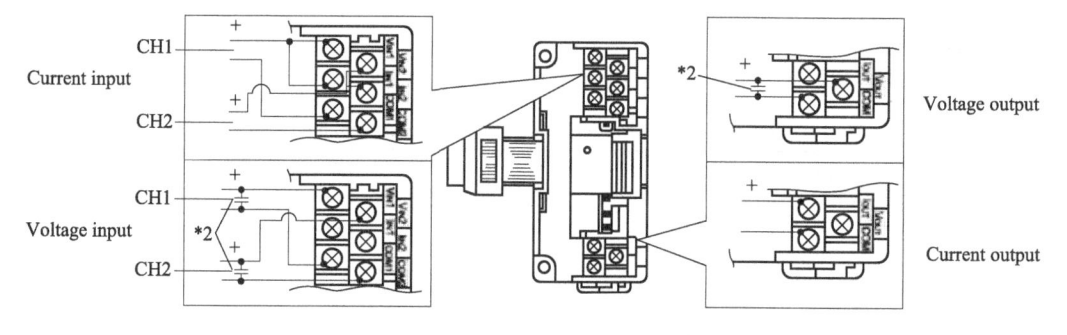

Figure 3. 4. 15   Wiring of FXON-3A module

When using current input, it shall make sure that the terminals marked with [VIN[*]1] and [IIN[*]1] are connected. When using current output, the [VOUT] and [IOUT] terminals shall not be connected. [*]1 here identifies terminal number 1 or 2.

If there is any voltage fluctuation or there is too much electrical noise in the voltage input/output, a capacitor with a rating of about 25V, 0. 1 to 0. 47μF shall be connected at position [*]2.

【Project Principle and Basic Knowledge】

According to aerodynamics, the kinetic energy of the wind is proportional to the square of the wind speed, and the wind power is proportional to the cube of the wind speed. When the wind speed is v, the input power P_{in} of the wind turbine can be expressed in the following formula:

$$P_{in} = \frac{1}{2}\rho A v^3 \tag{3.4.1}$$

$$A = \pi R_w^2 \tag{3.4.2}$$

In the formula, ρ is air density; A is the area swept by the wind turbine; v is the wind speed; R_w is the radius of the wind turbine;

If all wind energy through the swept surface of the impeller is absorbed by the fan blade, then the wind speed after passing through the impeller should be equal to zero, but the air can not be completely static, so the efficiency of the wind turbine is always less than 1, so the wind energy utilization coefficient of the wind turbine can be defined as C_p:

$$C_p = \frac{P_o}{P_{in}} \tag{3.4.3}$$

In the formula, P_o is the shaft power output by the fan.

Therefore, the shaft power P_o output by the fan can be expressed in the following formula:

$$P_o = \frac{1}{2}\rho A v^3 C_p \tag{3.4.4}$$

The wind energy utilization coefficient C_p (also called power coefficient) is an important parameter to characterize the efficiency of wind turbine. The C_p characteristic is usually obtained by experiments of wind turbine manufacturer. The C_p in this software is

given by statistical formula.

$$C_p = 0.5\left(\frac{RC_f}{\lambda} - 0.22\beta - 2\right)e^{-0.255\frac{RC_f}{\lambda}} \tag{3.4.5}$$

Of which, C_f is the design constant of the blade, usually being $1\sim3$. The pitch angle β is the angle between the velocity direction of the air flow and the string.

The factors influencing C_p are wind speed, blade speed, wind turbine radius and pitch angle, so the C_p characteristic of wind turbine is complex, but the maximum wind energy utilization coefficient C_{pmax} of wind turbine can be known by "Betz limit theory", C_{pmax} is approximately equal to 0.593.

In order to analyze C_p characteristic, another important parameter of wind turbine, blade tip velocity ratio λ, is defined, that is the ratio of blade tip linear velocity to wind speed, as shown below:

$$\lambda = \frac{R_w \omega_w}{v} = \frac{\pi n_w}{30 v} \tag{3.4.6}$$

In the formula, ω_w, blade rotation angular velocity, rad/s; n_w, blade speed, r/min.

The characteristics of the wind turbine are usually represented by a series of curves of wind energy utilization coefficient C_p, as shown in Figure 3.4.16 and Figure 3.4.17.

It can be seen from Figure 3.4.16 that the $C_p(\lambda)$ curve is a function of the pitch angle β, and when the β angle increases, the $C_p(\lambda)$ curve decreases significantly.

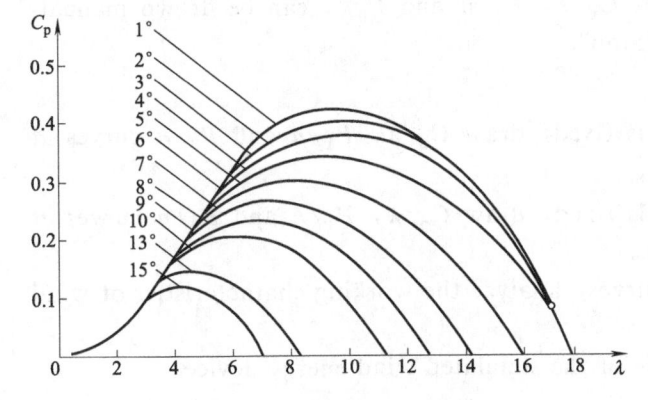

Figure 3.4.16　Characteristic curve of wind turbine (1)

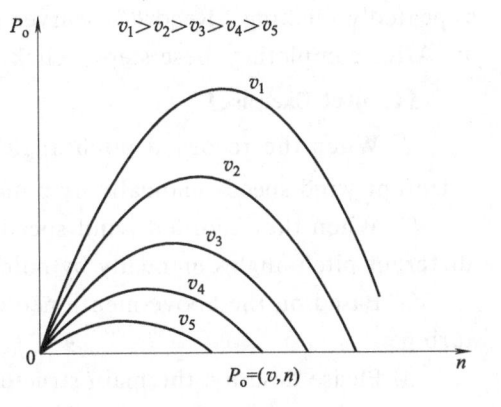

Figure 3.4.17　Characteristic curve of wind turbine (2)

The C_p characteristic is usually obtained by experiments of wind turbine manufacturer. The C_p characteristic curve in this software is given by statistical formula.

【Project Implementation】

(1) Curvilinear observation of wind energy utilization coefficient and speed ratio

① Double-click "Stand-alone Server" icon on the desktop to enter the monitor system, click "Login" and then click the "Wind Turbine Characteristic Simulation" button in the top window to enter the simulation interface. Set the real-time wind speed to 6 m/s, and the other parameters to the default values. Then click "Automatically draw the curve" and "Start", in the C_p/λ

Wind turbine characteristic simulation

diagram, the change curve of C_p with λ can be seen. In P_1/n, P_0/n diagrams, the change curve of input power, output power and torque of the fan with angular velocity can be observed. Click "Stop" after completion.

② Set the real-time wind speed to 12m/s, repeat, observe the differences between the curves of C_p/λ, P_1/n and P_0/n and these when the real-time wind speed is 6m/s.

(2) The influence of the pitch angle on the utilization coefficient of wind energy

① After setting the pitch angle to 5 degrees and repeating, observe the differences between the curves of C_p/λ, P_1/n and P_0/n and these when the pitch angle is 0 degree.

② After setting the pitch angle to 20 degrees and repeating, observe the differences between the curves of C_p/λ, P_1/n and P_0/n and these when the pitch angle is 0 and 5 degrees.

(3) Manually draw curves of C_p/λ, P_1/n, P_0/n by points

Enter the wind power monitoring software, and click the wind turbine characteristics simulation into the simulation interface. Real-time wind speed is set to 6m/s, and the other parameters are the default values. After clicking in turn "Draw the curve manually" and "Start", click "Confirmation" after inputting wind turbine speed in the wind turbine speed setting column, and then click "Record". In C_p/λ, P_1/n and P_0/n diagrams, the points corresponding to the speeds of the wind turbine can be observed. After inputting different wind turbine speeds in the wind turbine speed display numerical value soft keyboard, and repeatedly clicking "Record", curves of C_p/λ, P_1/n and P_0/n can be drawn manually. After completing these steps, click "Stop".

【Project Exercise】

① When the recorded pitch angle is fixed, draw C_p/λ, P_1/n and P_0/n curves at different wind speeds manually by points.

② When the recorded wind speed is fixed, draw C_p/λ, P_1/n and P_0/n curves at different pitch angles manually by points.

③ Based on the above-mentioned curves, analyze the working characteristics of wind turbine.

④ Please describe the main structure of the simulated wind energy device.

Project Ⅴ　Verification of Maximum Power Tracking Algorithm for Photovoltaic Array

【Project Description】

Grasp the MPPT algorithm programming, learn the software and hardware components of the energy conversion storage system, the basic operation of the software and the data acquisition operation, and be able to make the secondary development of the engineering project of the system based on the control signal and the collected data.

Section Ⅲ Practical Training Project

【Ability goals】

① To learn the maximum power tracking algorithm of photovoltaic array.

② To grasp the principle and application of the maximum power tracking algorithm of photovoltaic array.

【Project Environment】

Debugging of photovoltaic array maximum power tracking

To complete this training task, we should refer to the training platform equipment description manual of THWPFG-4 wind-solar complementary power generation system, study the electrical schematic diagram of the energy conversion storage control unit, the composition of the MPPT controller, the electrical index and interface description, and the principle of using the MPPT controller; understand the working principle of CPU core module and carry out program design and debugging of maximum power tracking according to the voltage and current sampling signal of PV array. We should understand and write different MPPT algorithm to achieve maximum power tracking and send adjusted parameters (duty cycle parameters) through the serial port to the PWM drive module to adjust. The microprocessor adopts 51 series single-chip microcomputer which has on-line download function and convenient programming debugging, can achieve MPPT control algorithm.

(1) MPPT controller

MPPT controller is mainly composed of DC voltage and current acquisition module 1, DC voltage and current acquisition module 2, CPU core module, human-computer interaction module, PWM driving module, communication interface module, temperature alarm module, DC-DC Boost main circuit module and intelligent charge and discharge controller, which can achieve the maximum power tracking of solar cells and improve the efficiency of solar cells effectively. It also improves the performance of the system.

The function modules of the MPPT controller (Figure 3.5.1) are explained as follows.

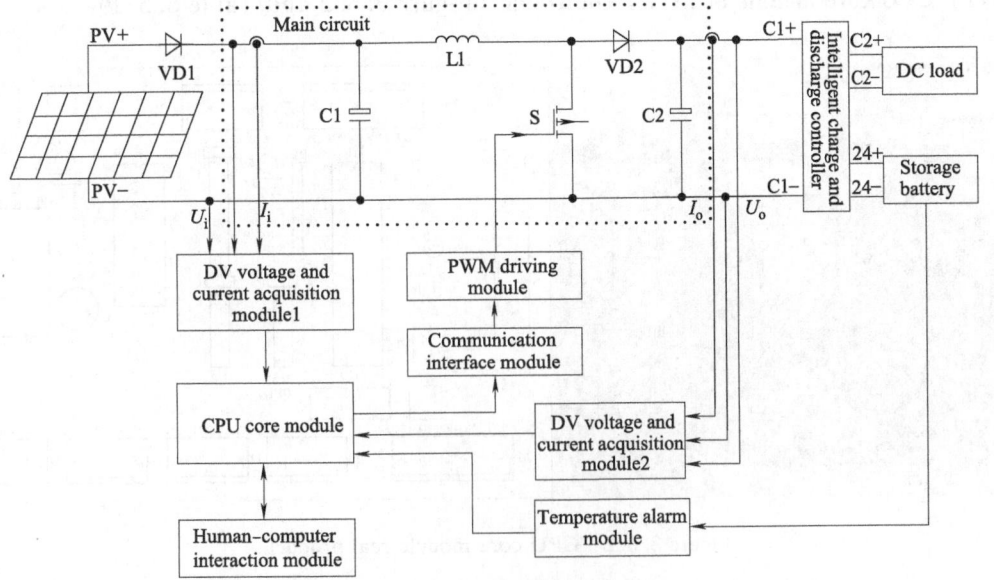

Figure 3.5.1 Function block diagram of MPPT controller

49

① DC voltage and current acquisition module 1: through the voltage Hall sensor and the current sensor, the voltage and the current output by the photovoltaic array battery are converted into voltage signals which meet the demand of input of single-chip microcomputer.

② Human-computer interaction module: input and output terminal of CPU core module.

③ Temperature alarm module: it detects the temperature of the battery. When the reference temperature is exceeded, the LCD will display that the temperature of the battery is too high.

④ PWM driving module: it receives the power regulation parameters (duty cycle parameters) output by the maximum power tracking microprocessor and outputs the PWM signals of different duty cycles. The PWM signal output by PWM micro-processing is isolated from DC/DC circuit, and the PWM signal is converted into the driving signal which satisfies the demand of the switch tube to improve the driving ability.

⑤ DC voltage and current acquisition module 2: through the voltage Hall sensor and the current sensor, the output voltage and current are converted into the voltage signal to meet the demand of single-chip microcomputer input.

⑥ Communication interface module: it mainly acts as signal level conversion to convert TTL level to 232 and 485 signals.

⑦ DC-DC Boost main circuit module: the duty cycle of the main circuit is adjusted by real-time collected power of the photovoltaic array cell. The impedance of the load is adjusted and the power used (namely the volt/sec area) is changed. It can always follow the maximum power point of the photovoltaic array cell output and it is equivalent to an impedance converter.

⑧ Intelligent charge and discharge controller: it adjusts the charging state and current according to the voltage of the battery to prevent the battery from overcharging or overdischarging and to prolong the battery life.

(2) CPU core module of MPPT controller (Figure 3.5.2 and Table 3.5.1)

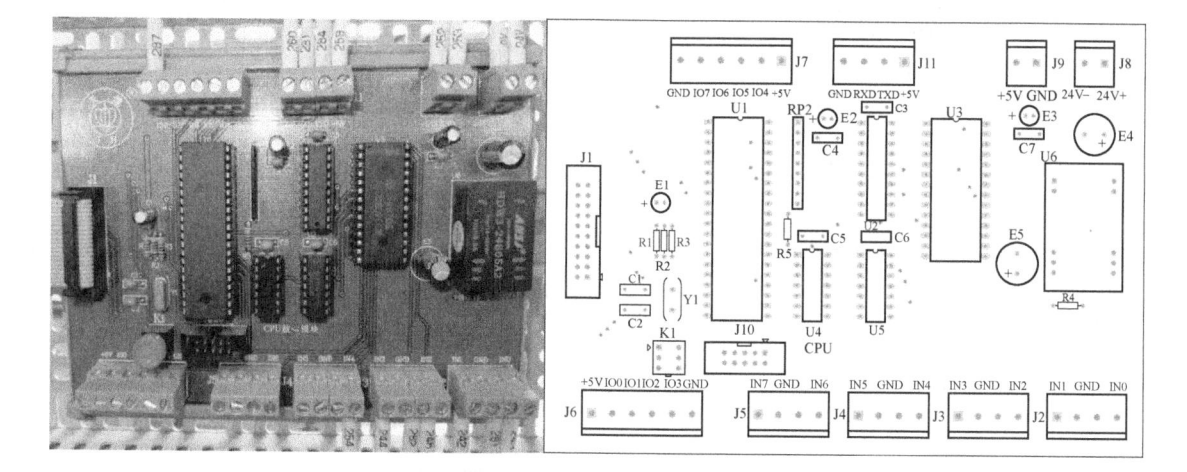

Figure 3.5.2　CPU core module real product

Section III   Practical Training Project

Table 3.5.1   Definition of CPU Core Module Port

S/N	Name	Description	Extended interface	Remark
1	J2:IN0	Voltage signal within 5V when the input of AD sampling is 0		
2	J2:GND			
3	J2:GND	Voltage signal within 5V when the input of AD sampling is 1		
4	J2:IN1			
5	J3:IN2	Voltage signal within 5V when the input of AD sampling is 2		
6	J3:GND			
7	J3:GND	Voltage signal within 5V when the input of AD sampling is 3		
8	J3:IN3			
9	J4:IN4	Voltage signal within 5V when the input of AD sampling is 4		
10	J4:GND			
11	J4:GND	Voltage signal within 5V when the input of AD sampling is 5		
12	J4:IN5			
13	J5:IN6	Voltage signal within 5V when the input of AD sampling is 6	√	
14	J5:GND		√	
15	J5:GND	Voltage signal within 5V when the input of AD sampling is 7	√	
16	J5:IN7		√	
17	J6:+5V	+5V power output	√	
18	J6:IO0	Voltage signal within 5V when switching volume I/O interface is 0	√	
19	J6:IO1	Voltage signal within 5V when switching volume I/O interface is 1	√	
20	J6:IO2	Voltage signal within 5V when switching volume I/O interface is 2	√	
21	J6:IO3	Voltage signal within 5V when switching volume I/O interface is 3	√	
22	J6:GND	Earth	√	
23	J7:+5V	+5V power output	√	
24	J7:IO4	Voltage signal within 5V when switching volume I/O interface is 4	√	
25	J7:IO5	Voltage signal within 5V when switching volume I/O interface is 5	√	
26	J7:IO6	Voltage signal within 5V when switching volume I/O interface is 6	√	

Continued

S/N	Name	Description	Extended interface	Remark
27	J7:IO7	Voltage signal within 5V when switching volume I/O interface is 7	√	
28	J7:GND	Earth	√	
29	J11: + 5V	+ 5V power output		
30	J11:TXD	Serial port sending end		
31	J11:RXD	Serial port receiving end		
32	J11:GND	Earth		
33	J9: + 5V	Isolation power supply DC/DC outputs 5V power positive pole	√	
34	J9:GND	Isolation power supply DC/DC outputs 5V power negative pole	√	
35	J8:24V +	Isolation power supply DC/DC inputs 24V positive pole		
36	J8:24V −	Isolation power supply DC/DC inputs 24V negative pole		
37	J1	To human-computer interaction module		20-row line

Programming precautions

① MPPT controller contains an intelligent charge and discharge controller. Because the intelligent charge and discharge controller needs at least 1 minute to finish the start after being powered on, it needs 1 minute delay when designing the maximum power trac-king program to ensure the intelligent charge and discharge controller work normally.

② The communication between CPU core module microprocessor and PWM driving module microprocessor adopts serial port communication. Baud rate is 9600; 8-bit data bit; one-bit stop bit; no check bit; one 8 Bit represents duty cycle, data range is 0x00～0xFF (0x00 means duty cycle is 0%; 0xFF means duty cycle is 99%)

③ The four-way stimulated signals to be converted are respectively connected to the IN0, IN1, IN2 and IN3 input channels of ADC0809: the assembly adopts accumulator A and external data memory to transmit instructions to select channels and read AD values Access address for each channel is as follows:

Address of PV array battery current (IN0) 0x7FF8;

Address of PV array battery voltage (IN1) 0x7FF9;

Address of Boost converter output current (IN2) 0x7FFA;

Address of Boost converter output voltage (IN3) 0x7FFB.

Using the assembly instruction to select the channel:

MOVX @ DPTR, A, where DPTR is the address, A can be any value [0x00～0xFF].

Using assembly instruction to read AD value:

MOVX A, @ DPTR, where DPTR is the address, A is AD conversion value.

④ The four-way stimulated signals to be converted are respectively connected to the

IN0, IN1, IN2 and IN3 input channels of ADC0809. Using the C language, each channel must be predefined as an external data area. The definition is as follows:

```
# define II_CURRENT XBYTE[0x7FF8]    /* PV array battery current */
# define UI_VOL XBYTE[0x7FF9]        /* PV array battery voltage */
# define IO_CURRENT XBYTE[0x7FFA]   /* Boost converter output current */
# define UO_VOL XBYTE[0x7FFB]        /* Boost converter output voltage */
```

Using C language to select the channel:

XXXX = X, where XXXX is a predefined name and X can be any value [0x00~0xFF].

Using C language to read the AD value:

the variable = XXXX, where XXXX is a predefined name and the content of the variable is the AD conversion value.

⑤ Example of programming (example of main program of constant volatage tracking)

```
UART_Init();                      //serial port initialization
duty= DUTY_DEFAULT;               //set default duty cycle parameter
uart_send_duty(duty);             //transmit duty cycle parameter to PWM microproces-
                                    sor
DelayMS(50);                      //delay
average_ui= average(0);           //acquisition of open circuit voltage value
v_const= (average_ui * 3)> 0 x02; //calculate reference voltage V_const= 0.75 * Voc
while(1)
{
  average_ui= average(0);         //input average voltage
  v_error= cabs(v_const-average_ui); //calculate the absolute error between the reference
                                    voltage and the input voltage
if(v_error> ERROR_VOLTAGE)        //the absolute error value is greater than the fixed
                                    value, adjust the duty cycle, otherwise the duty
                                    cycle is unchanged

  {
  if(v_const> average_ui)         //the reference voltage is greater than the input volt-
                                    age, the output voltage is reduced and the input
                                    voltage is increased

  {
    duty= Duty-DUTY_STEP;         //reduce duty cycle, reduce output voltage
  }
  else                            //the reference voltage is smaller than the input
                                    voltage, the output voltage is increased and the
                                    input voltage is reduced

  {
    duty= duty+ DUTY_STEP;        //increase the duty cycle, increase the output voltage
  }
  uart_send_duty(duty);           //transmit the duty cycle parameter to PWM micropro-
                                    cessor
DelayMS(50);                      //delay
    }
  }
```

(3) the human-computer interaction module of MPPT controller (Figure 3.5.3 and Table 3.5.2)

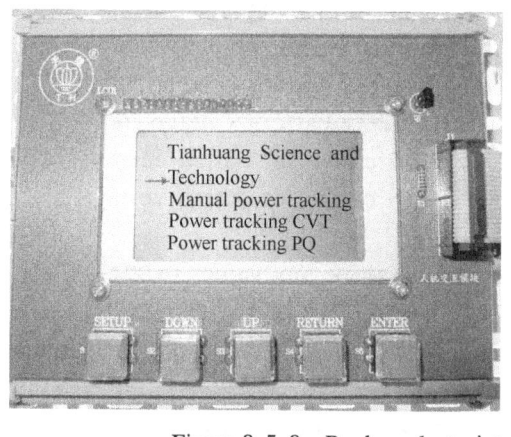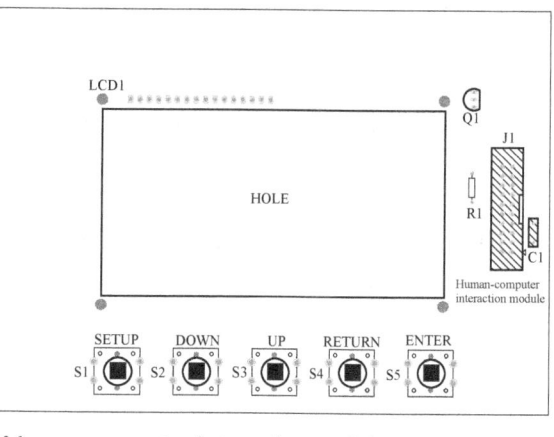

Figure 3.5.3 Real product picture of human-computer interaction module

Table 3.5.2 Definition of human-computer interaction module port

S/N	Name	Description	S/N	Name	Description
1	ENTER	Enter button	11	R/W	Read/write signal
2	RETURN	Return button	12	D7	Data Bit 7
3	INT1	Button common end	13	E	Enabling signal
4	UP	Up button	14	D6	Data Bit 6
5	GND (GND)	Earth	15	D0	Data Bit 0
6	DOWN	Down button	16	D5	Data Bit 5
7	VCC	Power	17	D1	Data Bit 1
8	SETUP	Setup button	18	D4	Data Bit 4
9	RS	Register selection	19	D2	Data bit 2
10	P10	Backlight control	20	D3	Data Bit 3

【Project Principle and Basic Knowledge】

(1) Constant voltage tracking (CVT)

Early control of the output power of photovoltaic cells was based on the technology of Constant Voltage Tracking (CVT). In Figure 3.5.4, L is the load characteristic curve, and when the temperature is maintained at a fixed value, it will have intersect points of a, b,

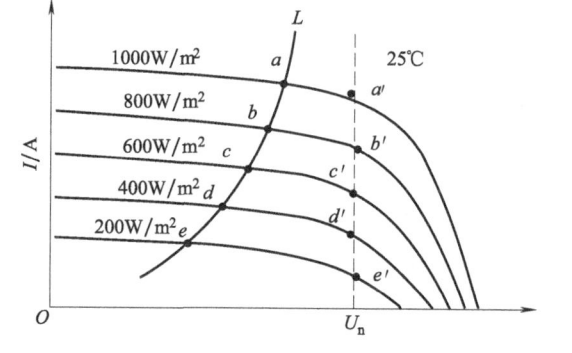

Figure 3.5.4 The volt-ampere characteristics of a silicon photovoltaic cell array

Section Ⅲ Practical Training Project

c, d and e with the volt-ampere characteristic curve at different sunshine intensity. It has been found that points of maximum power that the array can provide, such as a', b', c', d' and e', are located on adjacent sides of the same vertical line, which makes it possible to approximate the trajectory of the maximum power points as a vertical line of voltage $U =$ const, i. e. by keeping the voltage U at the output end of the array as a constant, the maximum power of the array can be roughly guaranteed at that temperature, so the maximum power point tracker can be simplified as a voltage regulator. This method is actually an approximate maximum power method.

CVT control has the advantages of simple control, high reliability, good stability and easy implementation. It is expected to get 20% more power than the general photovoltaic system. But the tracking method ignores the effect of temperature on the open-circuit voltage of solar cells. Taking monocrystalline silicon battery as an example, when the ambient temperature is increased by 10℃, the drop rate of its open circuit voltage is 0.35% ~ 0.45%. This shows that the voltage U_n at the maximum power point of the photovoltaic cell also varies with the ambient temperature. In areas where the seasonal temperature difference or diurnal temperature difference is large, the CVT control does not always track the maximum power in all temperature environments. The system control flow chart is shown in Figure 3.5.5.

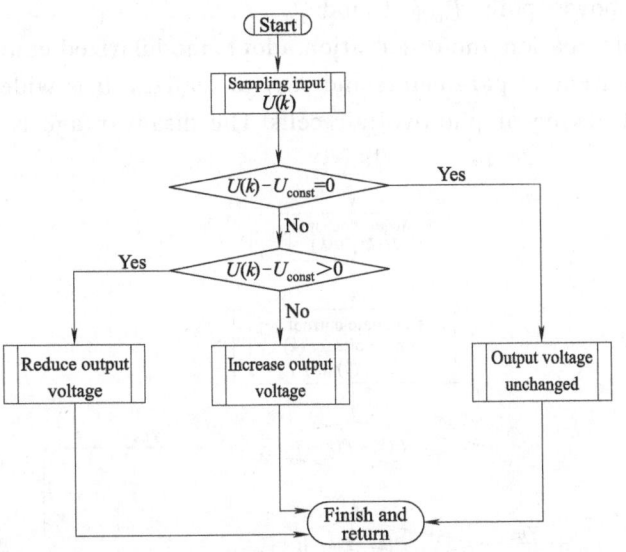

Figure 3.5.5 Control flow chart using CVT control

(2) Perturbation and Observation (P&O)

Perturbation and Observation (P&O) is also known as Hill Climbing (HC). The working principle is to measure the output power of the current array, and then to add a small voltage component to the original output voltage and the output power will be changed. By measuring the changed power and comparing it with the power before the change, the direction of the power change can be known. If the power is increased, the original disturbance is continued, and if the power is reduced, the direction of the original disturbance is changed, and the tracking situation of perturbation and observation is shown in Figure 3.5.6.

Supposing that the working point is at U_1 and the output power of the photovoltaic

55

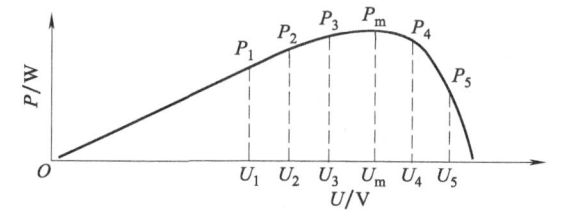

Figure 3.5.6　Tracking situation of perturbation and observation

cell is P_1, if the working point is moved to $U_2 = U_1 + \Delta U$ and the output power of the photovoltaic cell is P_2. Then, comparing the current power P_2 with the memory power P_1, if $P_2 > P_1$, it is shown that the input signal difference ΔU enlarges the output power, and the working point is located on the left side of the maximum power value P_m and the voltage needs to be increased continuously, so that the working point continues to move to the right, i. e. in the direction of P_m. If the working point has passed the P_m to U_4, then if the ΔU is added, then the working point reaches U_5, if the comparison result is $P_5 < P_4$, indicating that the working point is at the right of the maximum power value P_m, and then the direction of the input signal needs to be changed, that is, after the input signal subtracts the ΔU each time, the current power and the memory power are compared until the maximum power point P_m is found.

Because the perturbation and observation adopts modularized control loop with simple structure, less measurement parameters and easy to realize, it is widely used in the maximum power point tracking of photovoltaic cells. The disadvantage is that near the maxi-

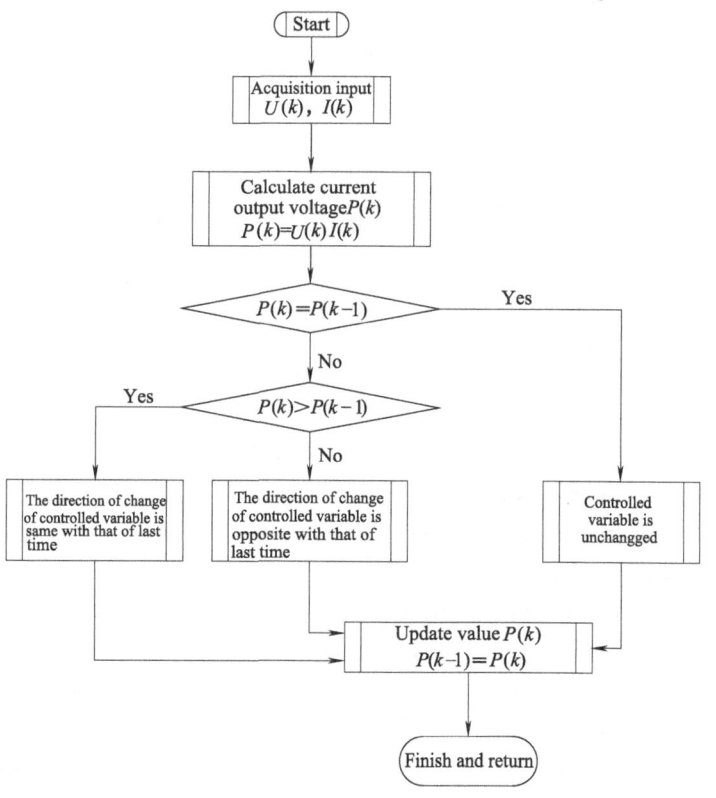

Figure 3.5.7　The Flow Chart of Perturbation and Observation Control

Section III   Practical Training Project

mum power point, the oscillation will be left and right resulting in energy loss, especially when the climate conditions change slowly, the situation is more serious. When climate conditions change slowly, the voltage and current of photovoltaic cells do not change much and this method will continue to disturb to change their voltage value and cause energy loss. Although the amplitude of each disturbance can be reduced to lower the oscillation amplitude of the point P_m to reduce the energy loss, this method will slow down the speed of tracking to another maximum power point when the temperature or illuminance changes significantly, which will waste a lot of energy. Therefore, the tracking step size, tracking accuracy and response speed can not be taken into account at the same time, and sometimes the program "misjudgement" phenomenon will occur in operation. The system control flow chart is shown in Figure 3.5.7.

Programming precautions and example of programming see P52.

【Project Implementation】

Photovoltaic array maximum power tracking algorithm

① Close the "main power supply" switch on the "stimulated energy control system", the system is electrified and the three-phase power indicator light is on.

② Close the air switch of the "switching power supply" to make the switching power supply work.

③ Close the air switch of the "simulated light source" to make the simulated sun lamp open.

④ Close the "PLC" air switch, power the PLC, pull the toggle switch on the PLC to the "RUN" state so that the PLC program is in the running state.

⑤ Press the "start" button (green button) of the sun tracking control switch to prepare the stepping motors and press the "control" button (yellow button) to simulate the light source to start the movement.

⑥ Close the "main power supply" switch on the "energy conversion and storage control system" and then the system is electrified and the three-phase power indicator light is on.

⑦ Record the PV output voltage that is the PV open-circuit voltage.

⑧ Place the brake of the charge and discharge controller in the "RELEASE (automatic brake)" state, close the air switch on the "storage battery" of the "energy conversion storage control system". The battery is connected to the PV MPPT control system and supplies the charge and discharge controller at the same time. At this time, the charge and discharge controller is initialized and the red indicator light is on (working in the braking state). The red light must be turned off (out of braking) before the next operation can be performed.

⑨ Close the "PV output" and "PV MPPT" air switch on the "energy conversion storage control system".

⑩ Press the reset button K1 on the "CPU core module" and then the system resets.

⑪ Press "ENTER" button on "human-computer interaction module" to enter the manual power tracking interface, then press "UP" and "DOWN" buttons to adjust the duty cycle manually, measure several groups of PV output current, PV output voltage, record the values and fill them in the appendix table 1 of this section.

⑫ By recording the voltage and current data, calculate the corresponding power of each voltage and current.

⑬ Through the "UP" and "DOWN" buttons on the "human-computer interaction module", choose "Power Tracking CVT" or "Power Tracking PQ", press "ENTER" button to enter the automatic power tracking mode. After a period of time, record the voltage and current data separately and calculate the corresponding power.

⑭ After the experiment is finished, all stepping motors stop running by pressing the "stop" button (red button). Then, turn off the air switches of the PLC, the switching power supply and the simulated light source and the main power supply switch on the stimulated energy control system, as well as the air switches of the photovoltaic MPPT and the of storage battery, and the main power supply switch on the the energy conversion storage control system. If a subsequent experiment will be carried out, the main power supply can not be turned off.

【Project Exercise】

① According to the measured data, draw the U-I curve and power curve, and find the maximum power point.

② Please calculate which point the maximum power output point is and why?

③ Please compare the maximum power points under automatic power tracking "CVT" and "PQ" and manual power tracking, and illustrate the advantages and disadvantages of the two automatic power tracking modes.

④ Please try to reprogramming the program to run and understand the maximum power tracking algorithm.

Project Ⅵ   Learn the Principle of Off-grid Inverter and Complete the System Test

【Project Description】

In order to accomplish this training task, we need to refer to the equipment instruction manual of training platform of THWPFG-4 wind-solar complementary power generation system and learn the circuit flow diagram of grid-connected inverter control unit, switch sequence of circuit breaker, and installation and use of off-grid inverter.

【Ability goals】

① To be capable of the function and usage of off-grid inverter.

② To be capable of the working principle of off-grid inverter.

【Project Environment】

The circuit flow diagram of the grid-connected inverter control unit is shown in Figure 3.6.1.

(1) Circuit breaker (Table 3.6.1)

Practice of off-grid photovoltaic inverter

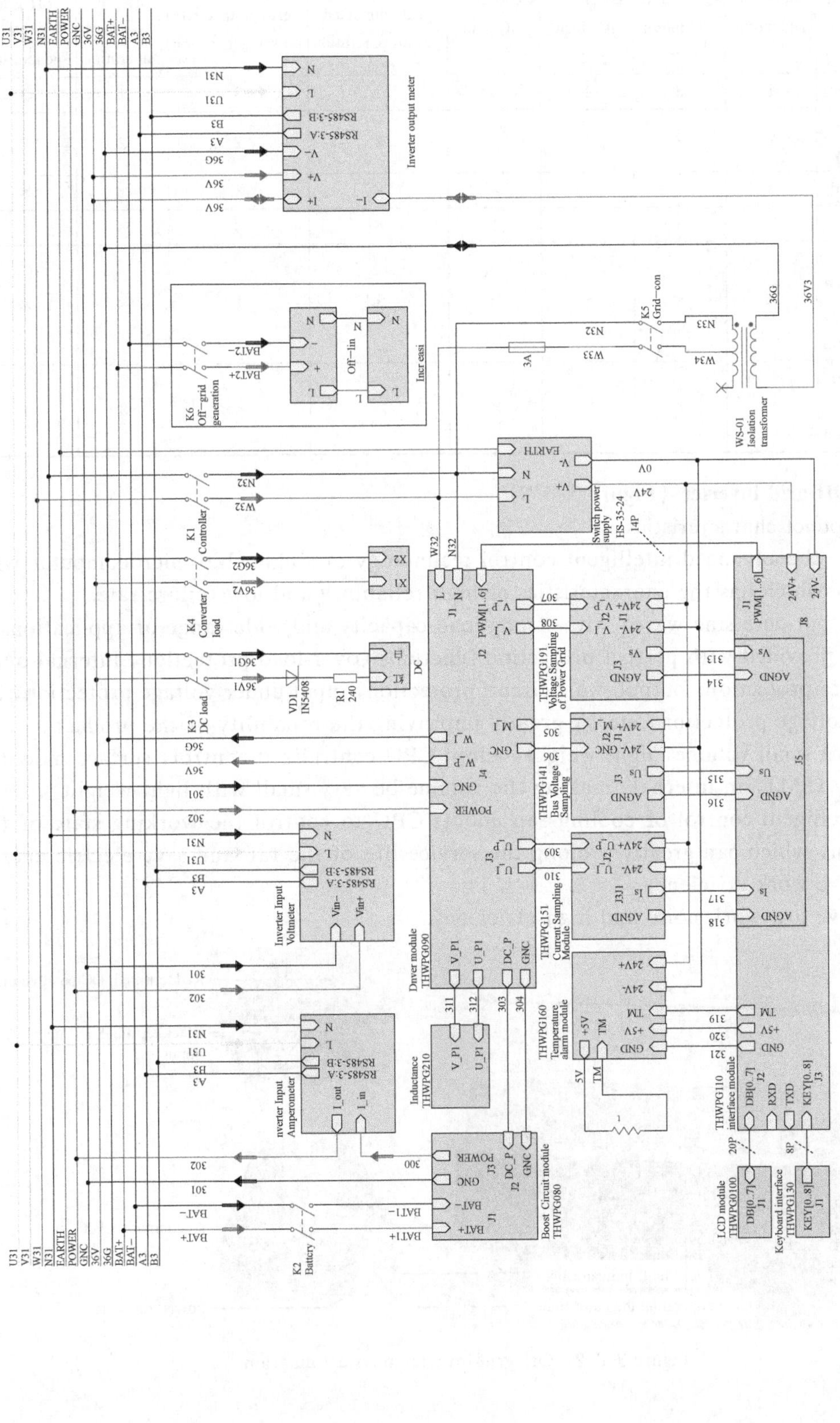

Section Ⅲ   Practical Training Project

Figure 3.6.1   Circuit flow diagram of grid-connected inverter control unit

59

Table 3.6.1 Switch sequence of circuit breaker

S/N	Function	Power-on sequence of DC motor	Power-off sequence of DC motor	Power-on sequence of AC load	Power-off sequence of AC load	Power-on sequence of grid-connected power generation	Power-off sequence of grid-connected power generation	Power-on sequence of off-grid power generation	Power-off sequence of off-grid power generation
K1	Controller	1	3	1	3	1	3	1	3
K2	Storage Battery	2	2	2	2	2	2	2	2
K3	DC load	3	1	×	×	×	×	×	×
K4	AC load	×	×	3	1	×	×	×	×
K5	Grid-connected power generation	×	×	×	×	3	1	×	×
K6	Off-grid power generation	×	×	×	×	×	×	3	1

(2) Off-grid inverter (Figure 3.6.2)

① Product characteristics

• Use the advanced intelligent control technology of single-chip microcomputer with dual-CPU, which has the characteristics of high reliability and low failure rate.

• Output pure sine wave, with strong load capacity and wide range of applications.

• Be provided with perfect protection functions (over-load protection, internal over-temperature protection, output short circuit protection, input under voltage protection, input over-voltage protection, etc.), greatly improving the reliability of the product.

• Own small volume, light weight, adopt CPU centralized control, surface mounted technology (SMT) in internal, making the volume be very small with light weight.

• Intelligent control of cooling fan adopts CPU to control the working state of the cooling fan, which can greatly prolong the service life of the fan and save electric energy and improve work efficiency.

• Have low work noise and high efficiency.

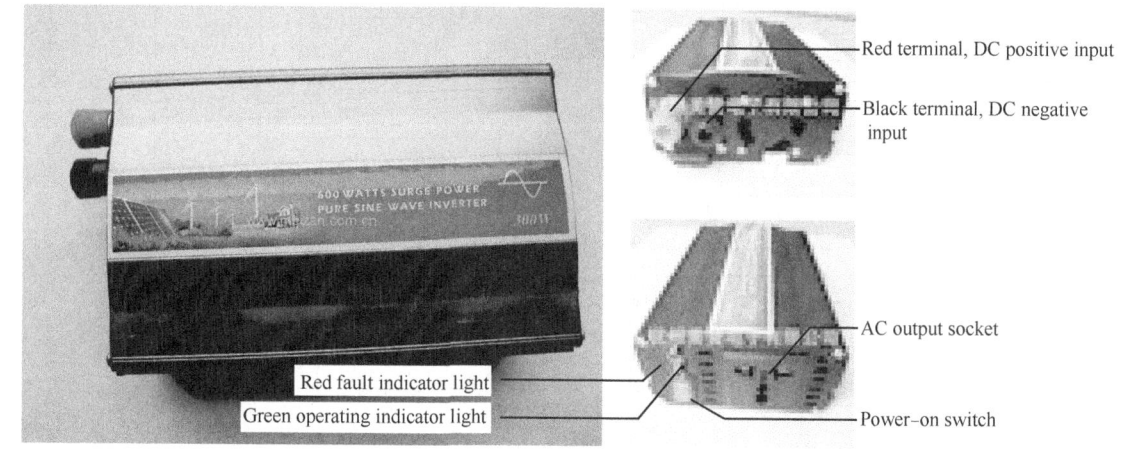

Figure 3.6.2 Off-grid inverter physical diagram

——— Section Ⅲ Practical Training Project

② Technical indicators （Table 3. 6. 2）

Table 3. 6. 2   Table of technical indicators of off-grid inverter

	Model	MZ-300W
Input	DC voltage	12V or 24V or 48V or 110V
	DC voltage range	10~15V DC or 21~30V DC or 42~60V DC or 100~120V DC
	No-load current	<0. 5
	Efficiency	>85%
	DC connection	Cables With Clips or Car Adaptor
Output	Output voltage	100/110/120V AC or 220/230/240V AC
	Rated power	300W
	Instantaneous power	600W
	Output waveform	Sine wave
	Output frequency	50Hz or 60Hz
	Waveform distortion	3%
Protection	Low voltage alarm	10V DC±0. 5V or 20. 5V DC±1V or 44V DC±1V or 100V DC±1V
	Under voltage lock out	9. 5V DC±0. 5V or 19. 5V DC±1V or 42V DC±1V or 96V DC±5V
	Overload	Output off
	Over voltage	15. 5V or 30. 5V or 61. 2V or 120V
	Over heat	Automatic output shut-down
	Fuse	Short circuit
Environment	Working temperature	Between −10℃ and +50℃
	Humidity	20%~90% RH without condensation
	Storage temperature	Between −30℃ and +70℃
Packaging	Machine size/mm	180×105×60
	Size including package/mm	245×122×70
	Net weight/kg	0. 78
	Gross weight/kg	1
	Way of packing	Color box packaging
Others	Start-up	Soft start
	Cooling method	Fan cooling
	Total waveform distortion	THD<5%

③ Method of installation and usage

• Power selection. Storage battery, automobile battery or power supply of solar power system should be adopted to ensure that the input voltage of 12V/24V/36V/48V/110V is suitable for this product. Inverter power supply is strictly forbidden to use the power supply higher than rated input voltage.

• Make the switch is on OFF (OFF) position (including inverter power supply and equipment), get the power from the battery, connect the inverter power supply with the input power supply to enable that the black wire connecting with the black terminal is negative, the red wire connecting with the red terminal is positive.

• The inverter power supply is connected with an electrical equipment to ensure that the load power of the inverter power supply is within the specified power, and that the maximum power of the inverter power supply should not be exceeded when the power is turned on. When the inverter power supply is connected with the device, turn on the inverter power supply and the switch on the device. It is normal for the red light to turn on and off repeatedly when starting, however, if it lasts for more than 1 minute, which should be considered as abnormal, please check if the load is too large.

④ Protection function

• Input under voltage protection. When the battery voltage is very low, the alarm gives an alarm, indicating that the DC supply voltage is reduced and the battery needs to be recharged. When the input voltage of 12V is less than $10V \pm 0.5V$, less than $20V \pm 1.0V$ for 24V, less than $30V \pm 1.0V$ for 36V, less than $40V \pm 2.0V$ for 48V, and less than $90V \pm 2.0V$ for 110V, the AC output will automatically turn off, and the indicator light and alarm will act simultaneously.

• Input over-voltage protection. When the input voltage reaches $15.5V \pm 0.5V$ for inverter power supply of 12V, reaches $31V \pm 1.0V$ for 24V, reaches $46.5V \pm 1.0V$ for 36V, reaches $62V \pm 2.0V$ for 48V, and reaches $132V \pm 2.0V$ for 110V, the indicator light turns red and the AC output automatically shuts down.

• Short circuit protection. When a short circuit occurs, the output device is turned off.

• Overload protection. When overload occurs, the output device will be turned off and the red light will be on.

• Opposite polarity input protection. When the battery is connected in reverse, the fuse burns to protect the device, and there is also the possibility of damage to the inside of the inverter power supply.

• Overheat protection. When the internal temperature exceeds about 75 degrees Celsius, the AC output automatically will shut down and the red indicator light is on. It can not be used until the temperature drops to normal temperature.

• Intelligent control of fan. When the internal temperature of the radiator exceeds 50 degrees Celsius, the inside fan will open automatically to cool the inverter power supply.

⑤ Common troubleshooting

• The inverter power supply doesn't react. Check if the contact between the battery and inverter power supply is good or not, re-connect.

• The output voltage is too low.

• Overload, namely, the load power exceeds the rated power. Turn off part of the equipment and turn on the inverter power supply again.

• Low input voltage. Ensure that the input voltage is within the rated range.

• Low voltage alarm.

• The battery is dead. Replace the battery.

• Low battery voltage or poor contact. Replace battery, check connection or clean connection terminal with dry cloth.

• The inverter power supply has no output.

• The battery voltage is too low. Replace the battery.

• Load capacity is too large. Shut down part of the equipment and restart the inverter power supply.

• Overheat protection of inverter power supply. Cool the inverter and place it in a ventilated place.

• The inverter power supply is failed to turn on. Restart the inverter power supply.

• The positive and negative opposite connects, and fuse is burnt. Replace the fuse with the one of same specification to reconnect again.

• The inverter power supply does not work. Check whether the power switch, fuse and battery connection are in good condition.

Section Ⅲ   Practical Training Project

【Project Principle and Basic Knowledge】

Usually, the process of converting AC to DC is called rectification, the circuit that completes the rectification is called the rectifying circuit, and the device that realizes the rectifying process is called the rectifying device or rectifier. Correspondingly, the process of converting DC energy into AC energy is called inverting, the circuit that completes the function of inverting is called inverter circuit, and the device that realizes the process of inverting is called inverting equipment or inverters.

Modern inverter technology is a science and technology to study the theory and application of inverter circuit, which is a practical technology based on industrial electronics technology, semiconductor device technology, modern control technology, modern power electronics technology, semiconductor converter technology, pulse width modulation (PWM) technology and so on. It mainly includes 3 parts which are semiconductor power integration device and its application, inverter circuit and inverter control technology.

(1) Classification of inverter

There are many kinds of inverters, which can be classified according to different methods.

① The inverter can be divided into power frequency inverter, intermediate frequency inverter and high frequency inverter according to the frequency of output AC energy. The frequency of power frequency inverter is the inverter of 50~60Hz; the frequency of the intermediate frequency inverter is usually 400Hz to more than 10kHz; High-frequency inverters typically have frequencies of up to a dozen kHz to MHz.

② The inverter can be divided into single-phase inverter, three-phase inverter and multi-phase inverter according to the phase number of output.

③ According to the whereabouts of power output of the inverter, it can be divided into active inverter and passive inverter. Inverters that transmit the power output from the inverter to the industrial power grid are called active inverters, while inverters that transmit the output power of the inverter to some kind of electric load are called passive inverter.

④ According to the type of inverter main power circuit, the inverter can be divided into single-ended inverter, push-pull inverter, half-bridge inverter and full-bridge inverter.

⑤ According to the type of inverter master switch, it can be divided into thyristor inverter, transistor inverter, field effect transistor inverter and insulated gate bipolar transistor (IGBT) inverter. It can also be summarized into two main categories: "semi-controlled type" inverter and "fully-controlled type" inverter. The reason for the name of "semi-controlled type" is that it does not have the ability of self-closing, and the component loses its control after the conduction, such as ordinary thyristors; And the reason for "fully-controlled type" is that it has the ability of self-switching, i. e. , the conducting and on-off of components can be controlled by the control electrode, such as the electric field effect transistor and the insulated gate bipolar transistor (IGBT).

⑥ Based on the DC power supply, it can be divided into voltage source inverter (VSI) and current source inverter (CSI). The former's DC voltage is near the constant and the output voltage is alternating square wave. The latter 's DC current is nearly the constant, the output current is alternating square wave.

⑦ In accordance with the output voltage or current waveform of the inverter, it can be divided into sine wave output inverter and non-sine wave output inverter.

⑧ In the light of the control mode of inverter, it can be divided into power frequency

modulation (PFM) inverter and pulse width modulation (PWM) inverter.

⑨ The inverter can be divided into resonant inverter, fixed-frequency hard-switching inverter and fixed-frequency soft-switching inverter by the working mode of the switching circuit of the inverters.

⑩ According to the inversion mode, it can be divided into load-commutation inverter and self-commutation inverter.

(2) Basic structure of inverter

The direct function of the inverter is to convert DC energy into AC energy. The core of the inverter device is the inverter switching circuit, which is referred to as the inverter circuit. This circuit completes the invert function through the on-off of the power electronic switch. The on-off of the power electronic switch device requires certain driving pulses, which may be adjusted by changing a voltage signal. A circuit that generates and regulates pulses, is usually called a control circuit or control loop. In addition to the above-mentioned inverter circuit and control circuit, there are protection circuit, output circuit, input circuit, output circuit and so on in the basic structure of the inverter.

(3) Working principle of inverter

The working principle of single-phase voltage full-controlled type PWM inverter is shown in Figure. 3. 6. 3. It is a common single-phase output full-bridge inverter main cir-

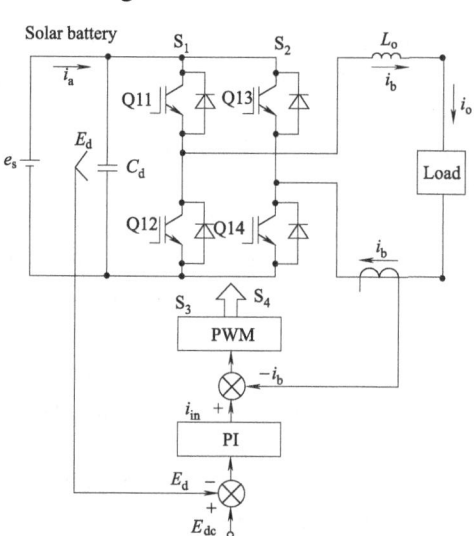

Figure 3. 6. 3 Single-phase Voltage PWM Inverter

cuit, it can be known form the figure that, the AC elements adopt IGBT tube Q11, Q12, Q13, Q14, and the conduction or cutoff of IGBT tube is controlled by by PWM.

After being connected with the DC power supply, the inverter circuit firstly is lead by Q11, Q14, stopped by Q12, Q13, then the current outputs by the positive pole of DC power, passes through Q11, L_o, load, transformer primary coil, then goes back to the negative pole of the power supply through Q14. When Q11, Q14 is off, it is lead by Q12, Q13, the current passes from the positive pole of the power supply through Q13, transformer primary coil, load, L_o, Q12, and returns to the negative pole of the power supply. At this point, positive and negative alternating square wave has formed on

the transformer primary coil, by using high-frequency PWM control, two pairs of IGBT tube carry out alternating conduction or cutoff to generate AC voltage in the transformer. Because of the LC AC filter, the output end is formed sine wave AC voltage.

【Project Implementation】

off-grid Inverter

① Turn on the main power switches of each control system; The power indicator light is shown.

② Put the brake for charging and discharging controller in the "RE-LEASE (Automatic Brake)" state and close the "storage battery" air switch of "energy conversion storage control system", power the charging and discharging controller, at this time the charging and discharging con-

Section III   Practical Training Project

troller initializes, the red indicator light is on (it works in the braking state). The red light must be turned off (out of braking state) before the next operation can be performed.

③ Open the "battery" and "off-grid power generation" switch of the "grid-connected inverter control system" in turn.

④ Start off-grid inverter (off-grid inverter installation should be reverse) to observe if off-grid load (fan) can run properly. Record the values of the battery voltmeter and the battery ammeter before and after the operation of the off-grid inverter (Table 3.6.3).

Table 3.6.3   Data recording

S/N	Items	Battery voltmeter	Battery ammeter
		U/V	I/A
1	Before working		
2	After working		

⑤ After ending the experiment, turn off "off-grid power generation" and "storage battery" air switches of "grid-connected inverter control system", and the "storage battery" air switch of the "energy conversion storage control system" by turn, and finally turn off the "main power" switch of each control system. It is not necessary to turn off the main power supply of the control system if there is subsequent experiments.

【Project Works】

① Refer to relevant data and describe the utility of off-grid inverter.

② If Q11 and Q14 of the single-phase voltage type PWM inverter are turned on, then which pole is the current output from, and through which switch back to the negative pole of the power supply?

Project VII   Test of Grid-connected Inverters and Analysis of Power Quality

【Project Description】

In order to accomplish this training task, we need to refer to the equipment instruction manual of training platform of THWPFG-4 wind-solar complementary power generation system and learn the composition of grid-connected inverter in grid-connected inverter control unit and parameter set of grid-connected inverter.

【Ability goals】

① To grasp the working principle of grid-connected inverter.

② To be capable of the usage and function of grid-connected inverter.

③ To grasp power quality and harmonic control.

【Project Environment】

(1) Composition of grid-connected inverter

The function flow diagram of grid-connected inverter is shown in Figure 3.7.1, the

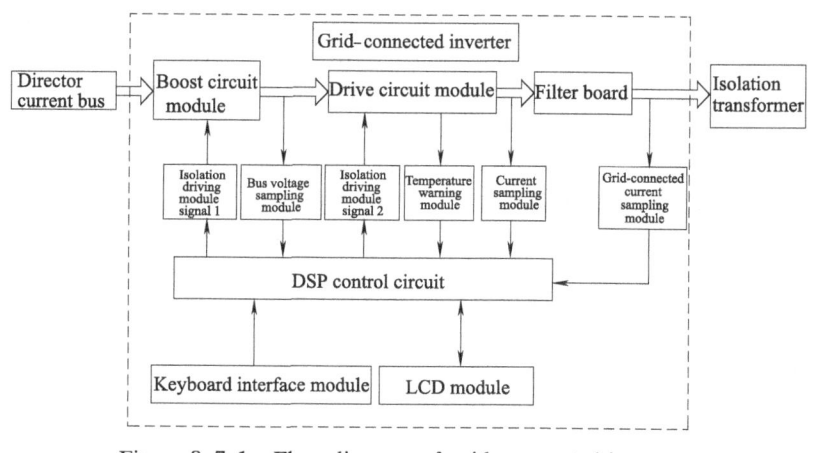

Figure 3. 7. 1　Flow diagram of grid-connected inverters

main circuit topology is composed of DC/DC (Boost circuit module) + DC/AC (drive circuit module) + filter (filter board), and the control loop is composed of bus voltage sampling module + current sampling module + grid voltage sampling module + temperature alarm module + isolation driving module + DSP control circuit + keyboard interface module + LCD module.

① The description of the grid-connected inverter module

• Boost circuit module: Boost circuit mainly converts DC voltage from DC bus output to bus voltage which can meet the requirements of grid connection.

• Drive circuit module: the drive circuit inverts the DC bus voltage through DC/AC to the sinusoidal alternating current which has the same frequency, phase and amplitude as grid voltage to realize the gid connection with power grid.

• Filter (filter board): filter out the high frequency PWM harmonic current output by the inverter, reduce the high frequency circulation in the incoming current and transfer the energy between the inverter and the power grid so that the grid-connected inverter can obtain certain damping characteristic so as to reduce the impulse current and be beneficial to the stable operation of the system.

• Bus voltage sampling module: detect the bus voltage and complete the function of closed-loop voltage and protection.

• Current sampling module: detect output current and complete the function of current closed-loop and protection.

• Power grid voltage sampling module: detect grid-connected voltage and complete the function of grid voltage phase-locked, voltage feedforward and protection.

• Isolation driving signal 1: complete the isolate driving function of Boost circuit module switch.

• Isolation driving signal 2: complete the isolate drive function of the drive circuit IPM intelligent module.

• Temperature alarm module: perform the temperature detection of IPM intelligent module, complete the over-temperature protection function of grid-connected inverter.

• DSP control circuit: execute the software algorithm function of grid-connected inverter.

• Keyboard interface module: set parameters which affect the quality of grid-connected current.

• LCD module: display grid-connected parameters.

② Drive circuit module (Figure 3. 7. 2, Figure 3. 7. 3 and Table 3. 7. 1)

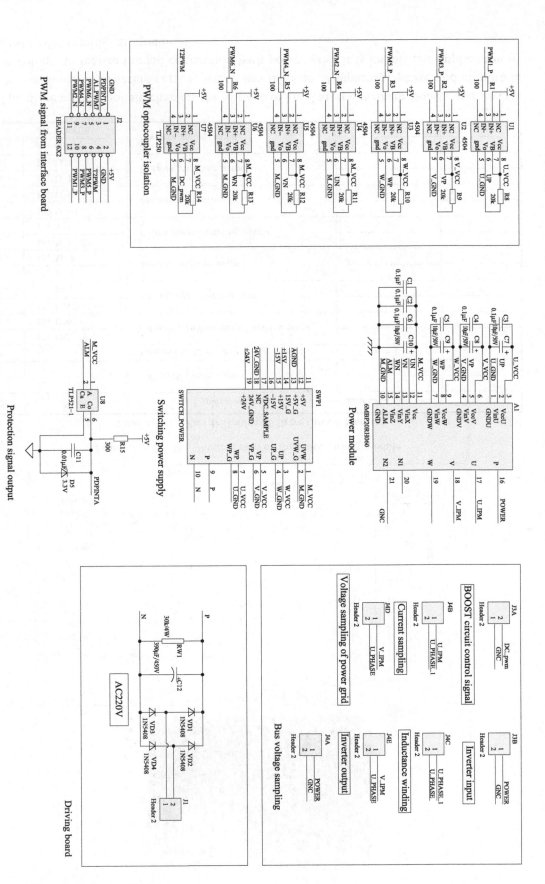

Figure 3.7.2 The schematic diagram of drive circuit module

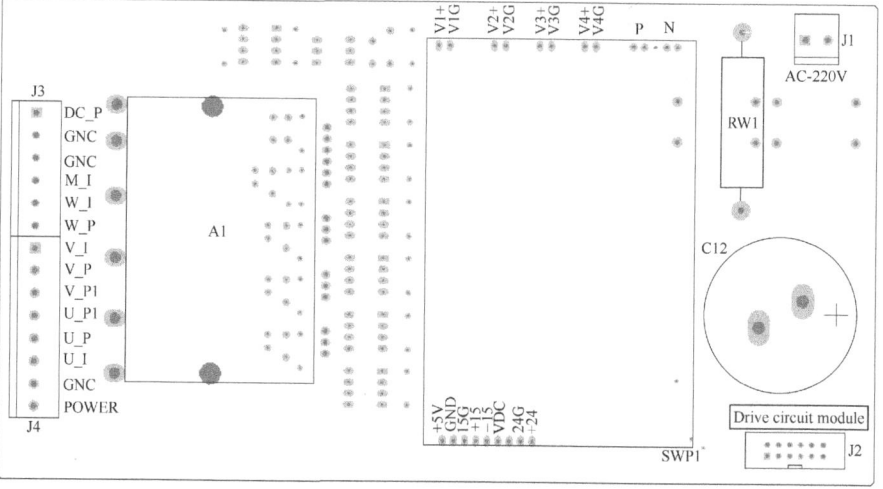

Figure 3.7.3　The physical map of the drive circuit module

Table 3.7.1　Port Definition of Drive Circuit Module

S/N	Name	Explanatory note	Extended interface	Remark
1	J1	AC 220V power supply		
2	J1			
3	J2	Connect to J1 of the interface board		
4	J3：DC _ P	Control signal output of voltage boost		
5	J3：GNC			
6	J3：GNC	DC voltage sampling output		
7	J3：M_I			
8	J3：W_I	Inverter voltage output		
9	J3：W_P			
10	J4：V_I	Grid voltage sampling output		
11	J4：V_P			
12	J4：V_P1	Inductance baffle		
13	J4：U_P1			
14	J4：U_P	Current sampling output		
15	J4：U_I			
16	J4：GNC	Inverter voltage input		
17	J4：POWER			

The drive circuit module changes the voltage of the battery through Boost, and inverts into the sinusoidal alternating current which has the same frequency, phase and amplitude as grid voltage to realize the gid connection with power grid. It is mainly the application of Mitsubishi IPM intelligent module.

Section Ⅲ   Practical Training Project

③ Boost circuit module（Figure3. 7. 4，Figure3. 7. 5 and Table3. 7. 2）

Boost circuit module mainly converts the 24V DC voltage output from the storage battery into the DC bus voltage which can meet the grid-connected requirements.

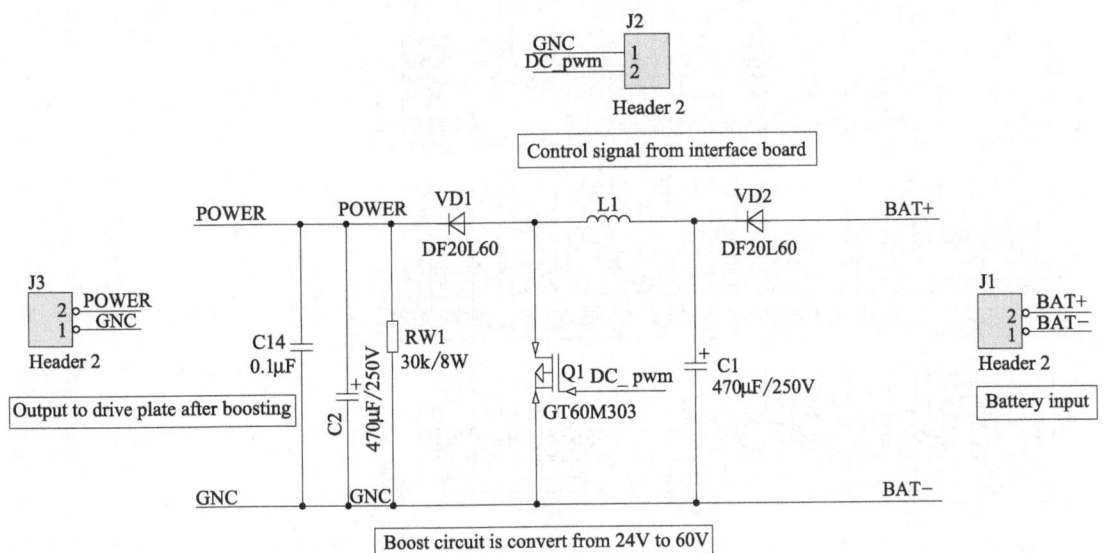

Figure 3. 7. 4   The schematic diagram of Boost circuit module

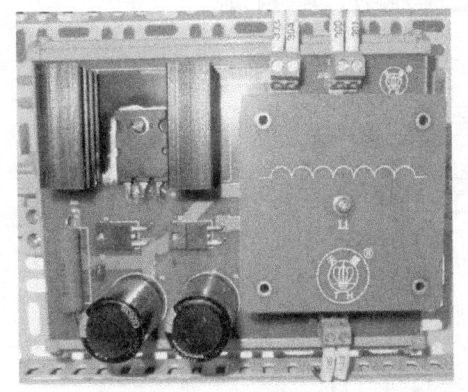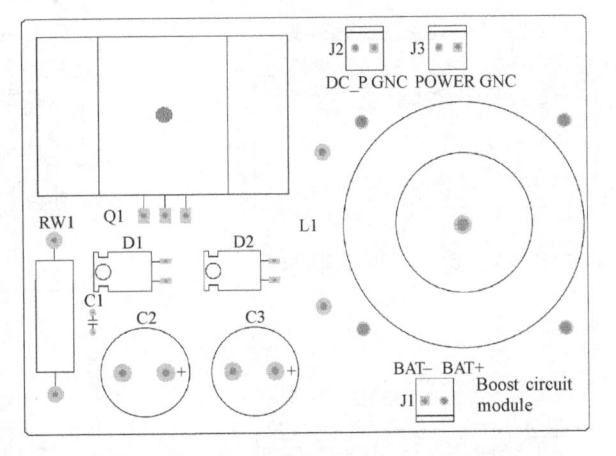

Figure 3. 7. 5   The physical drawing of Boost circuit module

Table 3. 7. 2   Port Definition of Boost Circuit Module

S/N	Name	Explanatory note	Extended interface	Remark
1	J1：BAT－	Battery 24V power input		
2	J1：BAT＋			
3	J2：DC＿P	Control signal input of voltage boost		
4	J2：GNC			
5	J3：POWER	60V output of voltage boost		
6	J3：GNC			

④ CPU module（Figure3. 7. 6，Figure3. 7. 7 and Table 3. 7. 3）

69

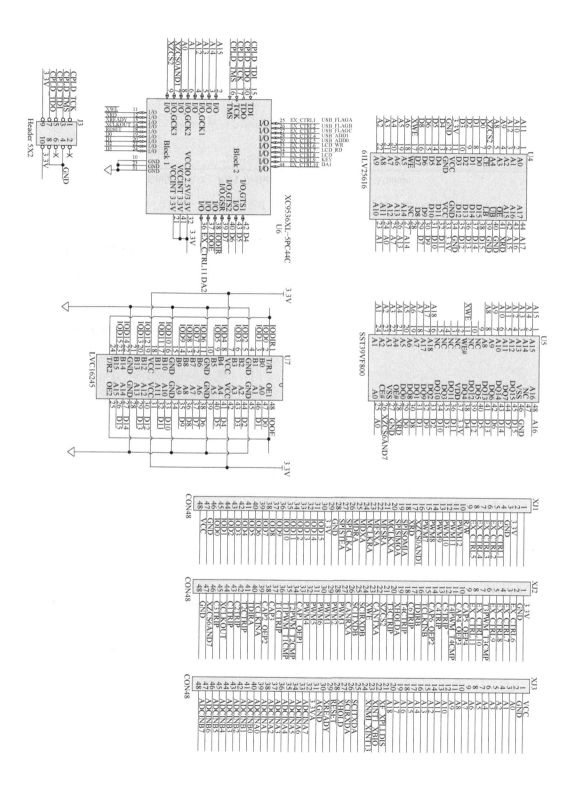

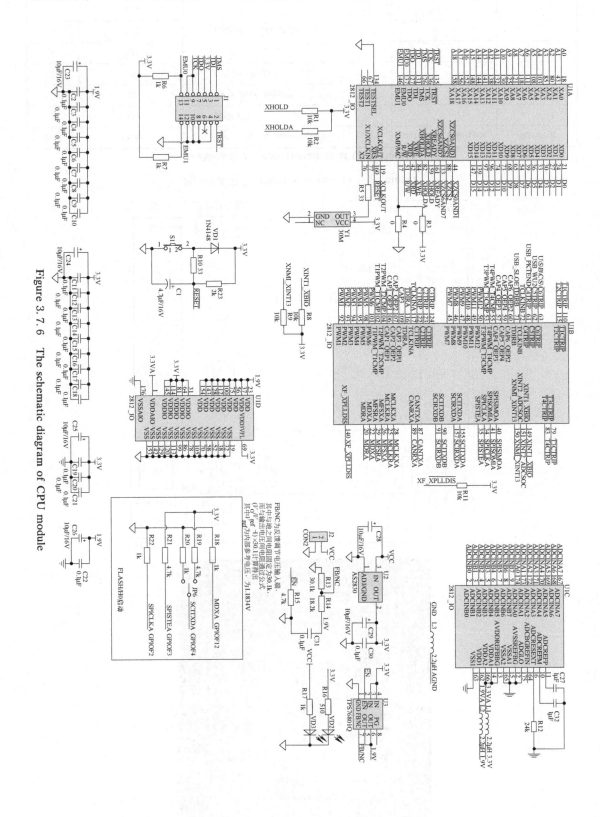

Figure 3.7.6 The schematic diagram of CPU module

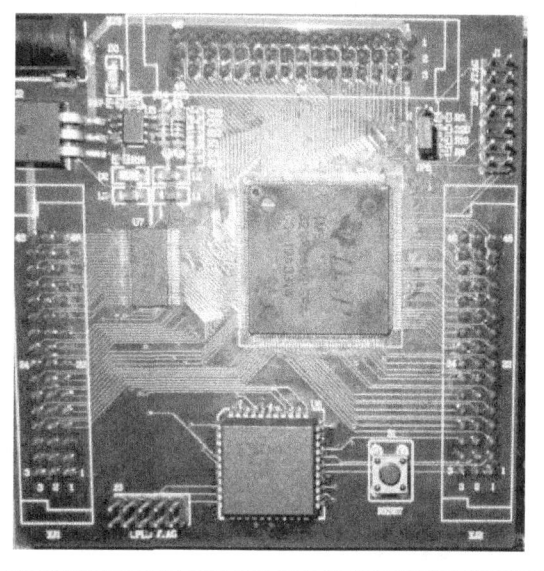

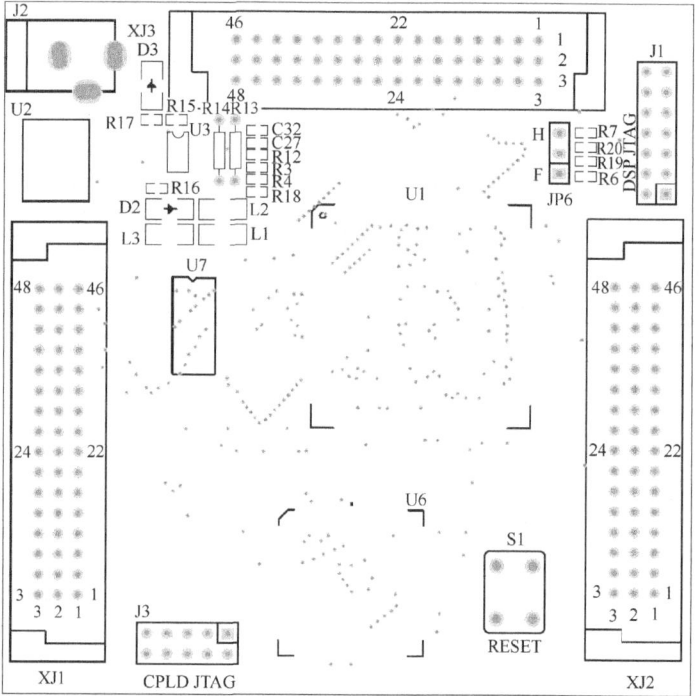

Figure 3. 7. 7　Physical photos of CPU module

Table 3. 7. 3　Port Definition of CPU Module

S/N	Name	Explanatory note	Extended interface	Remark
1	J1	DSP JTAG		
2	J2	5V Power Supply		For debugging
3	J3	CPLD JTAG		
4	XJ1			
5	XJ2	Plug in interface board		
6	XJ3			

Section Ⅲ Practical Training Project

The CPU module completes the software algorithm of grid-connected inverter. The CPU module adopts TI（Texas Instruments）company's TMS320LF2812 of 32-bit fixed-point DSP chip with high performance as the core chip.

⑤ LCD module（Figure 3.7.8，Figure 3.7.9 and Table 3.7.4）

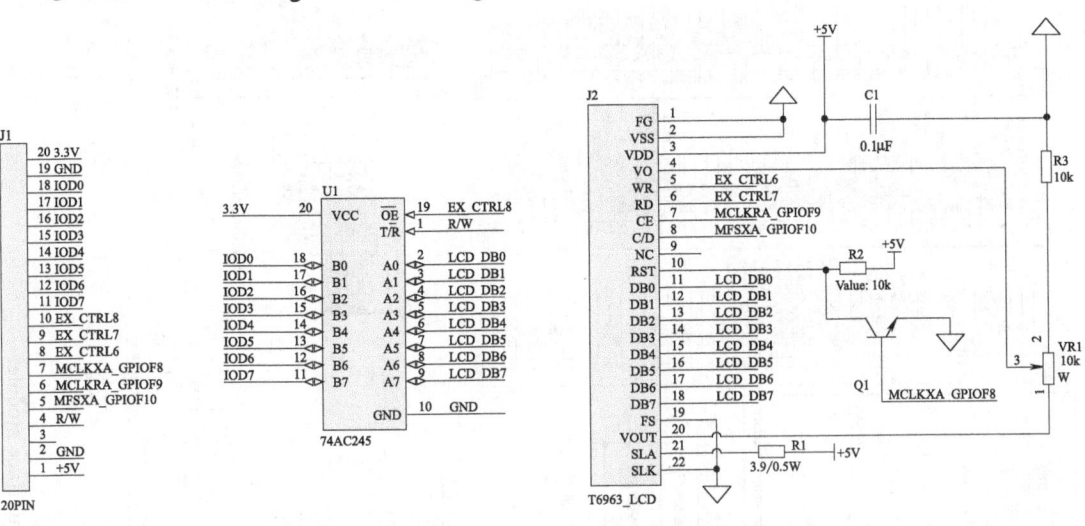

Figure 3.7.8　The schematic diagram of LCD module

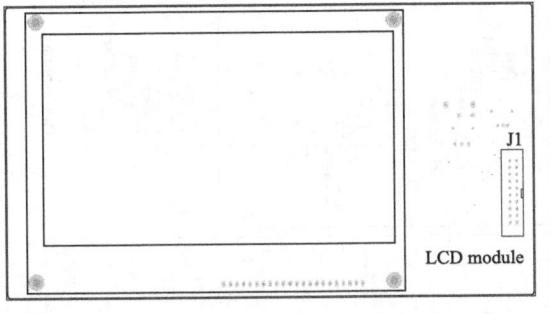

Figure 3.7.9　Physical drawing of LCD module

Table 3.7.4　Port Definition of LCD Module

S/N	Name	Explanatory note	S/N	Name	Explanatory note
1	+5V	5V Power Supply	11	IOD7	Data Bit 7
2	GND（GND）	Ground	12	IOD6	Data Bit 6
3	None		13	IOD5	Data Bit 5
4	R/W	Read/Write Signal	14	IOD4	Data Bit 4
5	GPIOF10		15	IOD3	Data Bit 3
6	GPIOF9		16	IOD2	data bit 2
7	GPIOF8		17	IOD1	Data Bit 1
8	EX _ CTRL6		18	IOD0	Data Bit 0
9	EX _ CTRL7		19	GND（GND）	Ground
10	EX _ CTRL8		20	3.3V	3.3V Power Supply

⑥ Interface module（Figure 3.7.10，Figure 3.7.11 and Table 3.7.5）

73

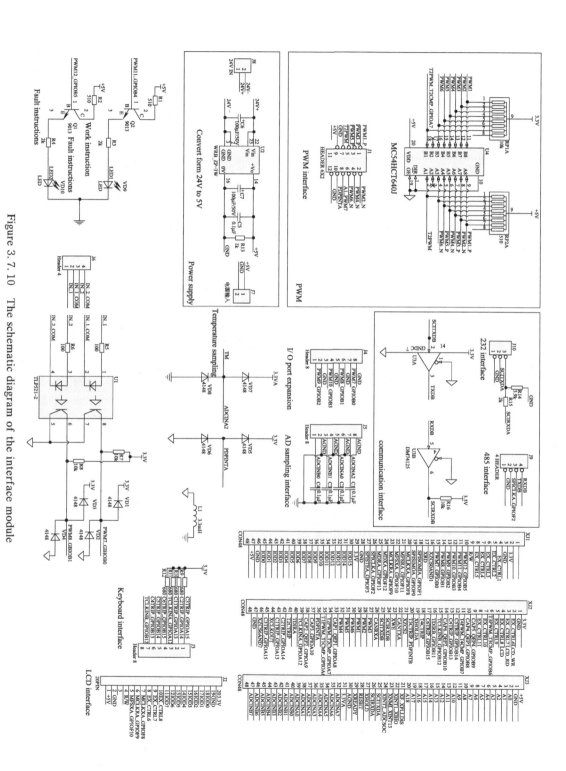

Figure 3.7.10 The schematic diagram of the interface module

Section Ⅲ   Practical Training Project

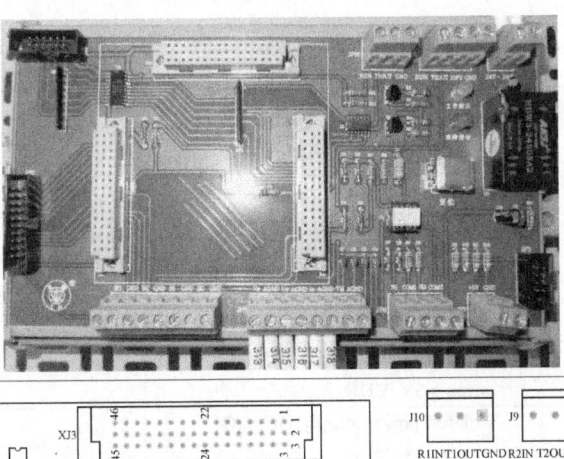

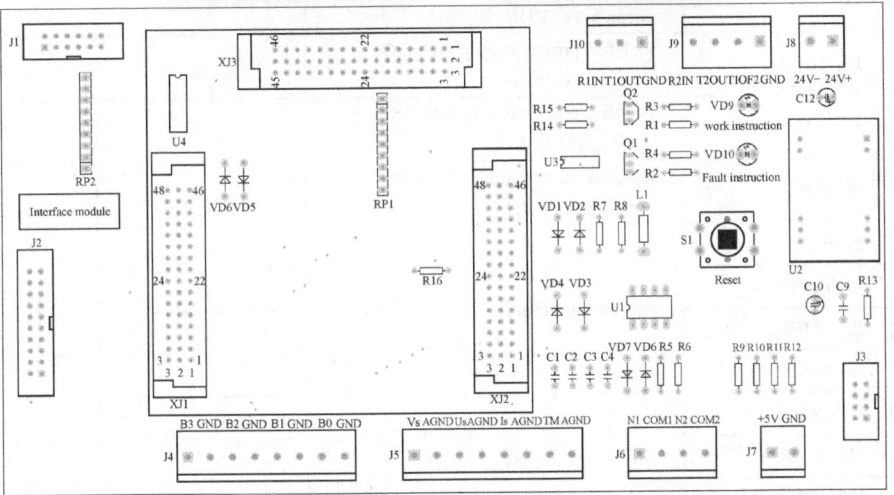

Figure 3.7.11   Physical drawing of interface module

Table 3.7.5   Port definitions of interface module

S/N	Name	Explanatory note	Extended interface	Remark
1	J1	To drive module	√	14P Flat Cable
2	J2	To LCD module	√	20P Flat Cable
3	J3	To keyboard interface module	√	8P Flat Cable
4	J4:B3	I/O port expansion	√	
5	J4: GND		√	
6	J4:B2	I/O port expansion	√	
7	J4: GND		√	
8	J4:B1	I/O port expansion	√	
9	J4: GND		√	
10	J4:B0	I/O port expansion	√	
11	J4: GND		√	
12	J5:Vs	DC voltage sampling input, voltage signal within 3.3V		
13	J5:AGND			
14	J5:Us	AC voltage sampling input, voltage signal within 3.3V		
15	J5:AGND			
16	J5:Is	Current sampling input, voltage signal within 3.3V		
17	J5:AGND			

75

Continued

S/N	Name	Explanatory note	Extended interface	Remark
18	J5:TM	Temperature sampling input, voltage signal within 3.3V		
19	J5:AGND			
20	J6:N1	I/O port expansion	√	
21	J6:COM1		√	
22	J6:N2	I/O port expansion	√	
23	J6:COM2		√	
24	J7: +5V	Isolated power supply DC/DC outputs 5V VDD		
25	J7: GND	Isolated power supply DC/DC outputs 5V VSS		
26	J8:24V −	Isolated power supply DC/DC inputs 24V VSS		
27	J8:24V +	Isolated power supply DC/DC inputs 24V VDD		
28	J9:R2IN	485 output signal	√	
29	J9:T2OUT		√	
30	J9: IOF2		√	
31	J9: GND		√	
32	J10:R1IN	232 output signal	√	
33	J10:T1OUT		√	
34	J10: GND		√	
35	XJ1	Insert core plate		
36	XJ2			
37	XJ3			

⑦ Keyboard interface module（Figure 3.7.12, Figure 3.7.13 and Table 3.7.6）
Set parameters which affect the quality of grid-connected current.

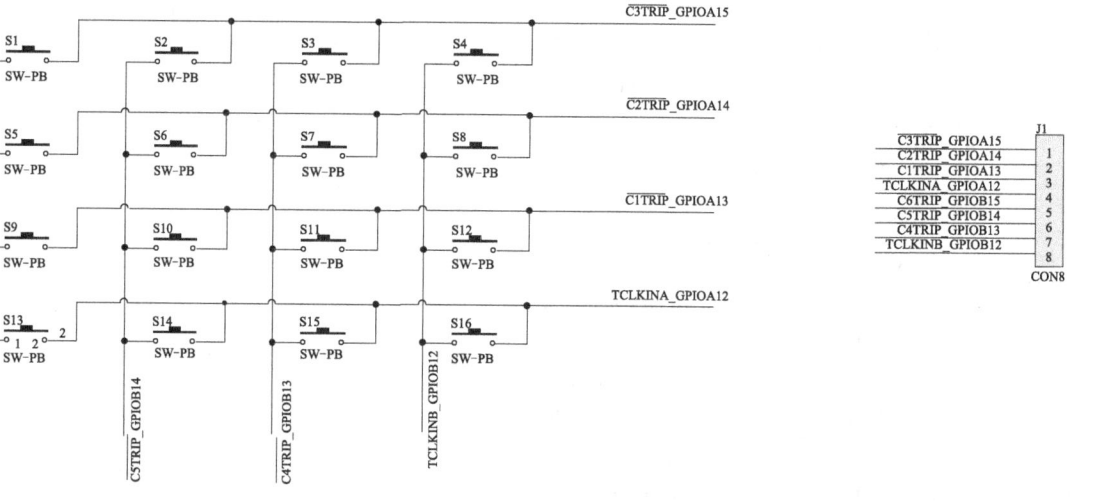

Figure 3.7.12　The schematic diagram of keyboard interface module

Section Ⅲ   Practical Training Project

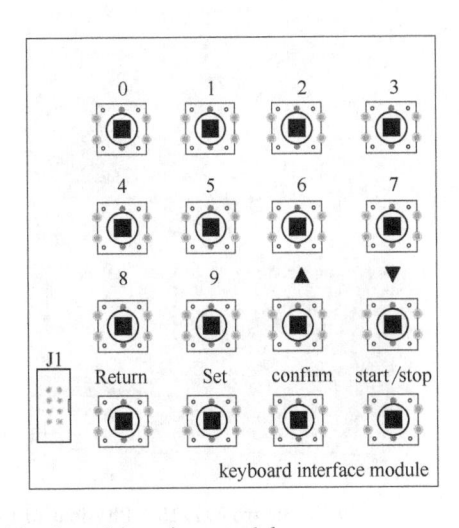

Figure 3. 7. 13   Physical drawings of keyboard interface module

Table 3. 7. 6   Port definition of keyboard interface module

S/N	Name	Explanatory note	Extended interface	Remark
1	J1	Connect to interface board J3		

⑧ Bus voltage sampling module（Figure 3. 7. 14，Figure 3. 7. 15 and Table 3. 7. 7）

The bus voltage sampling module completes the fucntion of voltage closed-loop and protection.

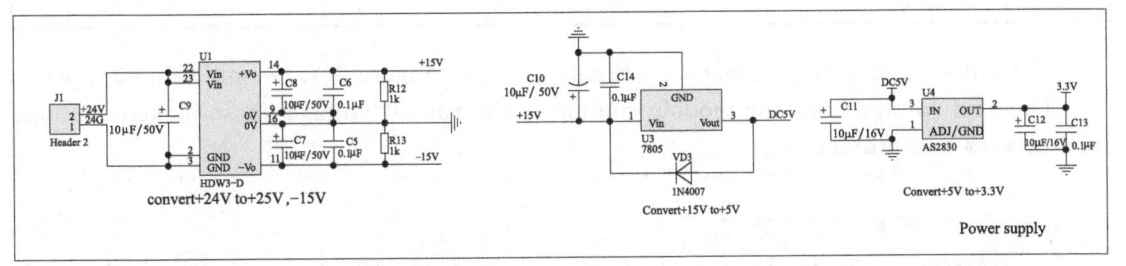

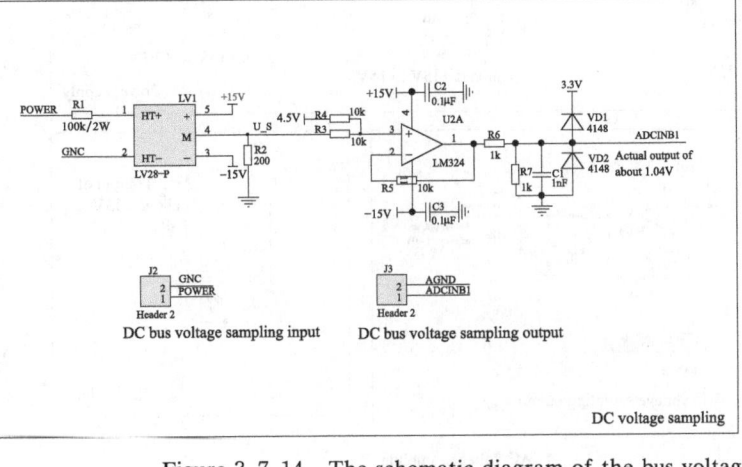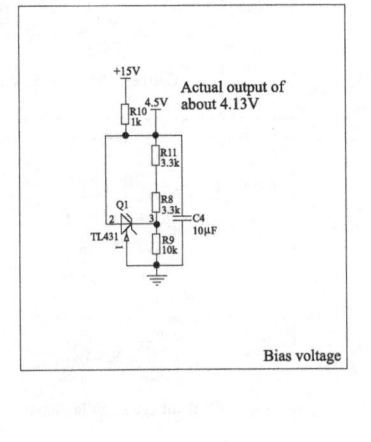

Figure 3. 7. 14   The schematic diagram of the bus voltage sampling module

77

Installation and Commissioning of Wind-solar Complementary Power Generating System

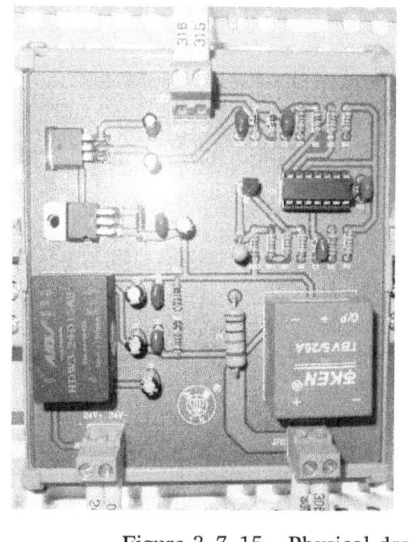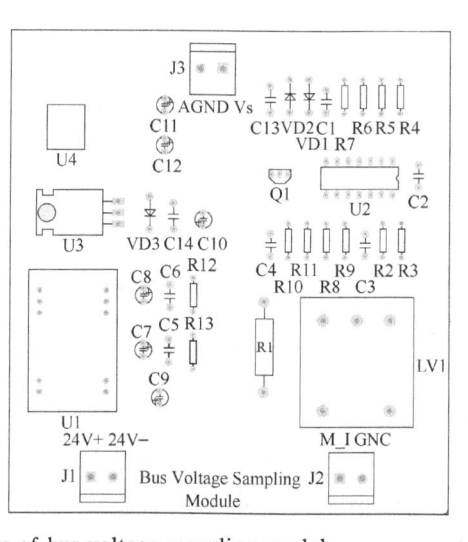

Figure 3.7.15　Physical drawings of bus voltage sampling module

Table 3.7.7　Port definitions of bus voltage sampling module

S/N	Name	Explanatory note	Extended interface	Remark
1	J1:24V +	Isolated power DC/DC 24V input		
2	J1:24V −			
3	J2:M_I	DC voltage sampling input		
4	J2: GNC			
5	J3:Vs	Voltage sampling control output		
6	J3:AGND			

⑨ Grid voltage sampling module（Figure 3.7.16，Figure 3.7.17 and Table 3.7.8）

The grid voltage sampling module completes the role of voltage phase-locked，voltage feedforward and protection.

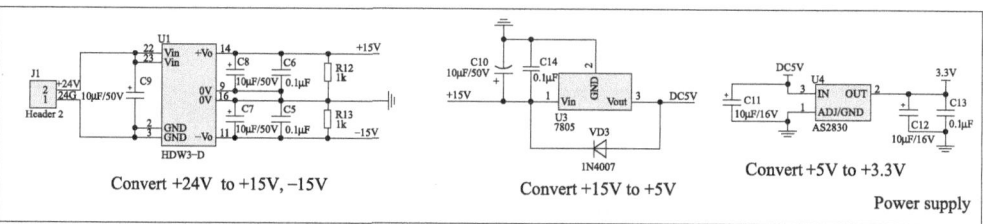

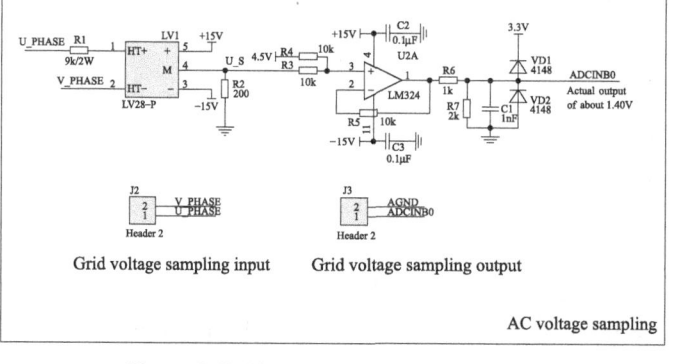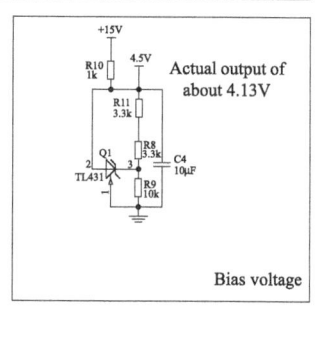

Figure 3.7.16　The schematic diagram of grid voltage sampling module

78

Section Ⅲ　Practical Training Project

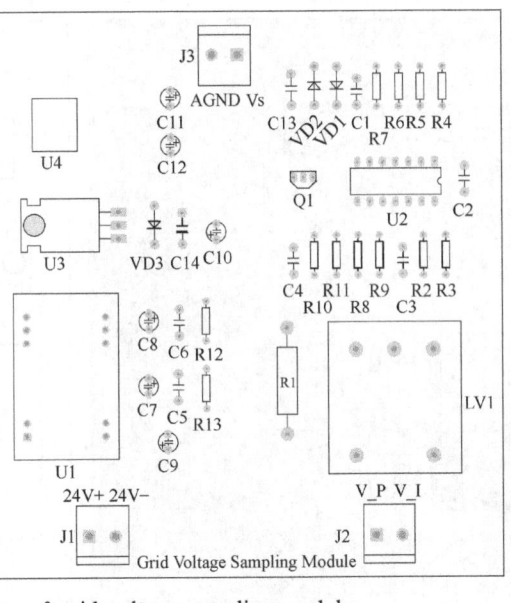

Figure 3.7.17　Physical drawings of grid voltage sampling module

Table 3.7.8　Port definition of grid voltage sampling module port

S/N	Name	Explanatory note	Extended interface	Remark
1	J1:24V +	Isolate power DC/DC 24V input		
2	J1:24V −			
3	J2:V_P	AC voltage sampling input		
4	J2:V_I			
5	J3:Vs	Voltage sampling control output		
6	J3:AGND			

⑩ Current sampling module（Figure 3.7.18，Figure 3.7.19 and Table 3.7.9）
Current sampling module completes the function of current closed-loop and protection.

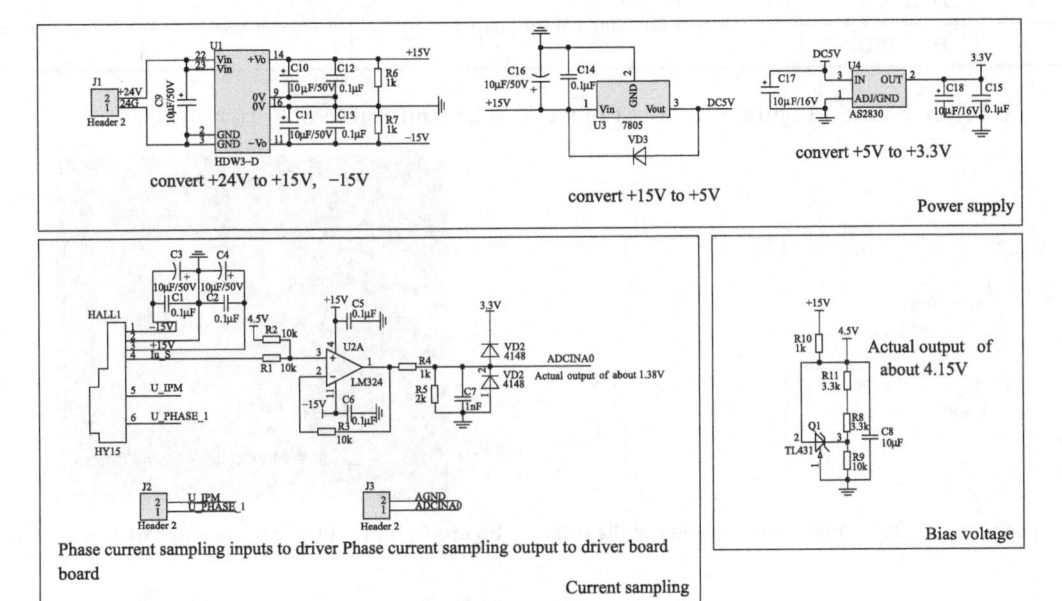

Figure 3.7.18　The schematic diagram of current sampling module

79

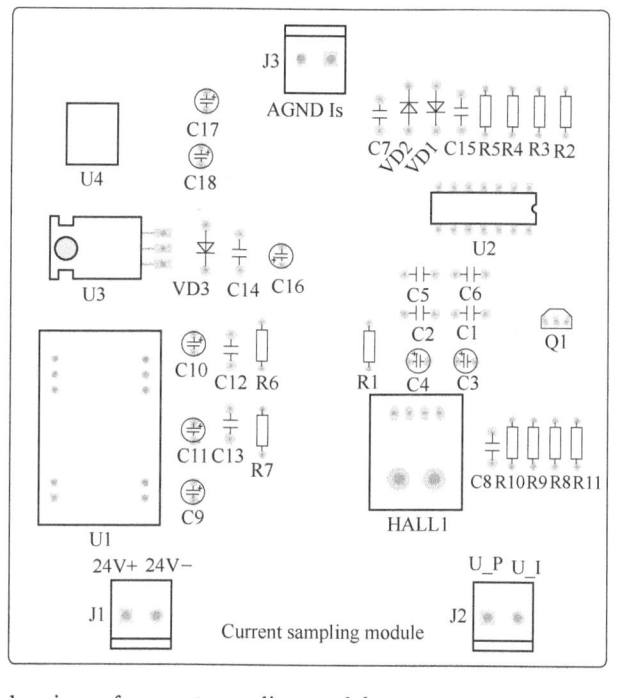

Figure 3.7.19　Physical drawings of current sampling module

Table 3.7.9　Port Definition of Current Sampling Module

S/N	Name	Explanatory note	Extended interface	Remark
1	J1:24V+	Isolated power DC/DC 24V input		
2	J1:24V-			
3	J2:U_P	Current sampling input		
4	J2:U_I			
5	J3:Is	Current sampling control output		
6	J3:AGND			

⑪ Filter board （Figure 3.7.20，Figure 3.7.21 and Table 3.7.10）

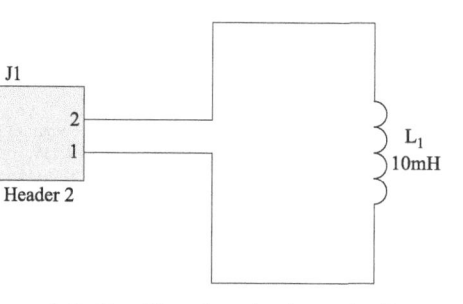

Figure 3.7.20　Filter board schematic diagram

Figure 3.7.21　Physical drawings of filter board

Section Ⅲ   Practical Training Project

The filter board is a filter comprised of inductive coil, and its main function is to filter out the high-frequency PWM harmonic current output by the inverter, reduce the high-frequency circulation in the incoming current and transfer the energy between the inverter and the grid so that the grid-connected inverter can obtain certain damping characteristic so as to reduce the impact current and be beneficial to the stable operation of the system.

Table 3.7.10   Filter board port definition

S/N	Name	Explanatory note	Extended interface	Remark
1	J1	Filter inductance		
2	J1			

(2) Operation instruction for parameter setting of grid-connected inverter

The controller starts to power, the system initialization, the interface module working indicator light and the fault indicator light are all bright, the LCD display "initializing", and after 3~5 seconds, the indicator light is out and the LCD screen shows the interface as shown in Figure 3.7.22.

Grid-connected Inverter

① Parameter setting step

• Click the "enter" button on the keyboard and there is a parameter setting interface, as shown in Figure 3.7.23.

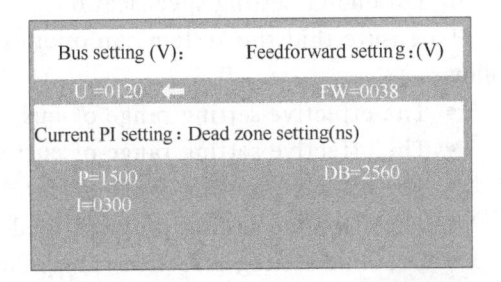

Figure 3.7.22   System default interface   Figure 3.7.23   Parameter Setting Interface

• With the help of the "▼" or "▲" key on the keyboard, the cursor on the LCD screen moves, select the parameters to be set, default to choose bus setting, click "set" key, the interface is shown as Figure 3.7.24.

• Set the desired value by the numeric keys on the keyboard, such as enter "1", "0" and "0", display U = 100, click "enter", and complete the parameter setting, as shown in Figure 3.7.25.

Parameter settings of grid-connected inverter

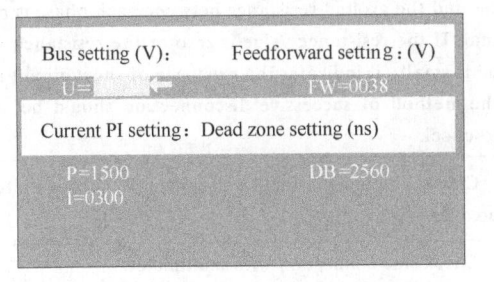
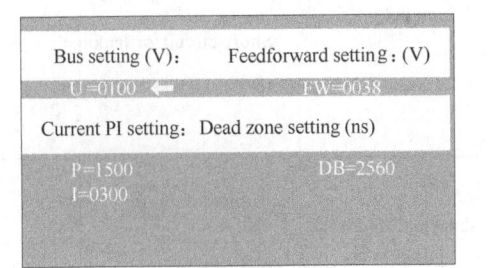

Figure 3.7.24   Parameter setting interface (1)   Figure 3.7.25   Parameter setting interface (2)

81

- Other parameter settings are similar to bus settings and operate as shown in Figure 3. 7. 26.

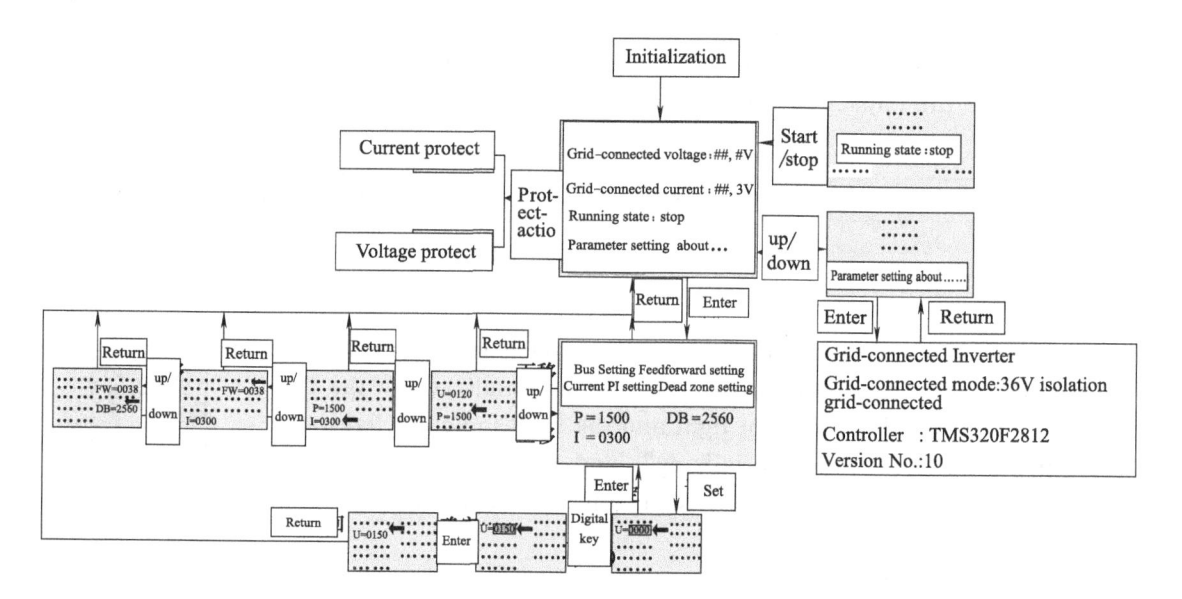

Figure 3. 7. 26   Functional diagram of operation panel

② Parameter setting specification

To ensure that the system can operate safely, the parameter ranges are defined as follows:

- The effective setting range of bus voltage: 60~120V;
- The effective setting range of current loop ratio coefficient P: 150~1500;
- The effective setting range of current loop integral coefficient I: 30~300;
- The effective setting range of feed-forward voltage: 0~50V;
- Deadline setting effective value: 2560 ns, 2780 ns, 2990 ns, 3200 ns.

(3) **Common trouble shooting (Table 3. 7. 11)**

Table 3. 7. 11   Common Troubleshooting Table

Faults symptom	Possible causes	Troubleshooting methods and procedures
Abnormal power supply	Fuse damage	Check whether the fuses in the fuse holder are burnt or not and replace a new fuse if the fuse is burned.
	Short circuit or leakage	Check whether there is a short circuit in the line and measure whether the resistance between different phases and the ground resistance between each phase is normal. If the resistance value is zero or the resistance value is small, it indicates the existence of short circuit and the method of successive disconnection should be used to check
Abnormal inverter power-on	Breaker damage	Check whether the input and output of DC load circuit breaker is on.
	Bad line contact	Check line connection with multimeter.
	The circuit breaker is not closed	Close circuit breaker

Section Ⅲ Practical Training Project

Continued

Faults symptom	Possible causes	Troubleshooting methods and procedures
After closing the grid-connected circuit breaker, the voltmeter is not shown	Breaker damage	Check whether the input and output of DC load circuit breaker is on.
	Bad line contact	Check line connection with multimeter.
	The circuit breaker is not closed	Close circuit breaker
	Without installing fuse core	Load the 3A fuse core
	Fuse core failure	Replace the fuse
Abnormal inverter working	Battery circuit breaker damage	Check whether the battery circuit breaker input and output are on.
	Battery circuit breakers not closed	Close the battery circuit breaker.
	Without starting-up inverter	Press the "start/stop" key of the keyboard interface module to start the inverter.
	Abnormal bus voltage sampling module line	Use a multimeter to check the bus voltage sampling module line connection, and test the voltage between Us and AGND at about 1.04V.
	Abnormal grid voltage sampling module line	Check the grid voltage sampling module line connection with the multimeter, and test the voltage between vs and AGND at about 1.38V.
	Abnormal current sampling module line	Use a multimeter to check the current sampling module connection and test the voltage between Is and AGND at about 1.38V.
	Abnormal LCD module	Check if the flat cable is loose and damaged
	Abnormal keyboard interface module	
Abnormal DC load	Breaker damage	Check whether the input and output of DC load circuit breaker is on.
	Bad line contact	Check line connection with multimeter.
	The circuit breaker is not closed	Close the circuit breaker
Abnormal AC load working	Breaker damage	Check whether the input and output of AC load circuit breaker is on.
	Bad line contact	Check line connection with multimeter.
	The circuit breaker is not closed	Close the circuit breaker

【Project Principle and Basic Knowledge】

(1) Control mode of inverter

According to the control mode, grid-connected inverter can be divided into voltage source voltage control, voltage source current control, current source voltage control and current source current control. However, due to the large inductance in the inverter circuit, the dynamic response of the system is often poor. At present, most of the inverters in the world are mainly voltage source input.

The output control run by the inverter and grid-connected can be divided into voltage control and current control. The power supply system can be regarded as a AC voltage source of constant value with infinite capacity. If the output of the grid-connected inverter is controlled by voltage control, it is actually a system running in parallel with the voltage source, in which, it is difficult to ensure the stable operation of the system. If the output of grid-connected inverter is controlled by current, the purpose of parallel operation can be achieved only by controlling the output current of grid-connected inverter to track the mains voltage.

The grid-connected inverter is usually controlled by voltage source input and current source output.

(2) Work principle of Inverter

The working principle of single-phase voltage full-controlled type PWM inverter is shown in Figure. 3. 7. 27. It is a common single-phase output full-bridge inverter main circuit, it can be known form the figure that, the AC elements adopt IGBT tube Q11, Q12, Q13, Q14, and the conduction or cutoff of IGBT tube is controlled by by PWM.

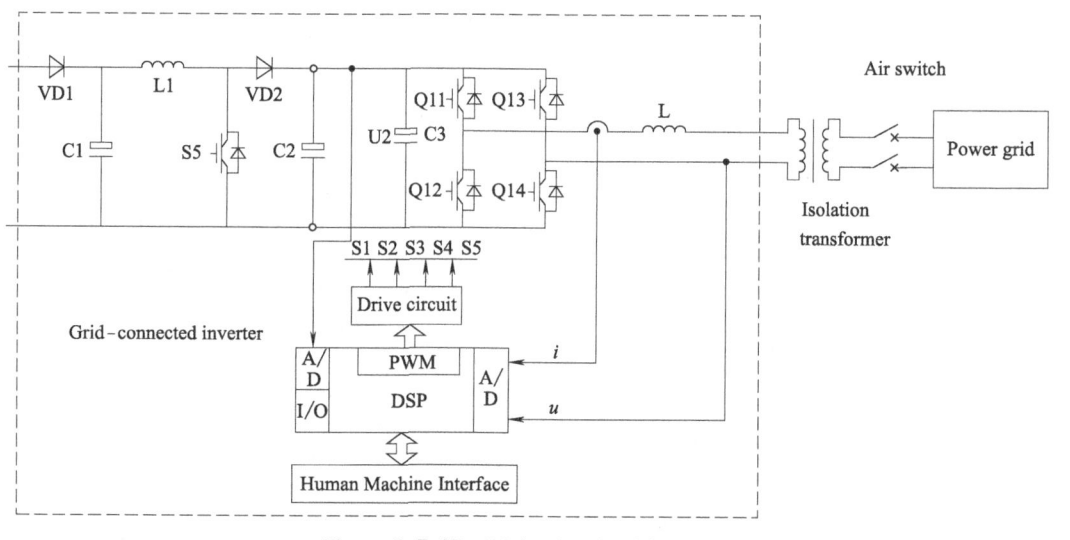

Figure 3. 7. 27　Main circuit of inverter

After being connected with the DC power supply, the inverter circuit firstly is lead by Q11, Q14, stopped by Q12, Q13, then the current outputs by the positive pole of DC power, passes through Q11, L, load, transformer primary coil, then goes back to the negative pole of the power supply through Q14. When Q11, Q14 is off, it is lead by Q12, Q13, the current passes from the positive pole of the power supply through Q13, transformer primary coil, Q12, and returns to the negative pole of the power supply. At this point, positive and negative alternating square wave has formed on the transformer primary coil, by using high-frequency PWM control, two pairs of IGBT tube carry out alternating conduction or cutoff to generate AC voltage in the transformer. Because of the LC AC filter, the output end is formed sine wave AC voltage.

With the development and application of new energy, the research of grid-connected inverter, a grid-connected device of distributed power generation system, is attracting more and more attention, and grid-connected inverter is a conversion device which can convert the DC power generated by power generation system to the AC power suitable for power grid. However, due to the increasing number of grid connected inverter devices put into use, the pollution of the incoming current harmonic output to the power grid can not be ignored. The total harmonic distortion rate (THD) of incoming current is usually used to describe the power quality. In addition to the additional loss, the harmonics in the incoming current may damage the powered devices of the power grid. The grid-connected inverter is controlled by voltage PWM technology to realize phase lock, DC bus regulation and current regulation of power grid so as to realize power grid-connection and load operation, with sine wave current output and unit power factor characteristics, effectively solving the

—————————————————— Section Ⅲ   Practical Training Project

harmonic pollution and low power factor of the grid-connected device.

(3) Current control strategy

By using the voltage outer loop and current inner control, the mathematical model of inverter in two -phase synchronous rotating reference frame is established firstly, based on which, the current closed-loop control strategy based on space vector modulation is proposed, realizing the independent control of active and reactive components of current in parallel network. The current inner loop is used to control the phase and amplitude of the grid-connected current, that is, to track the current command i_{ref}. The current inner loop uses the PI controller added to the grid voltage feedforward. The sampling value of the current feedback and the current command value are compared. The error is obtained through the PI controller. The output instruction type is the voltage. After adding with the feedforward voltage of the network, the required grid-connected voltage instruction is obtained. In fact, the addtion of grid voltage feedforward counteracts the grid voltage so that the output value of PI current loop is inductor voltage, thus fine-tuning the grid-connected current, which is a kind of advanced control. In the voltage feedforward PI control, the steady-state error is inevitable when the PI controller is tracking sinusoidal signal, so the actual output current can not be the same as the output current command.

There are two kinds of current inner loop control: current hysteresis tracking control and constant switching frequency current control, and this device adopts the latter, and the principle diagram is shown in Figure 3. 7. 28. Comparing the sine current reference value i_{ref} with the output Instantaneous Current i_o, the error value is adjusted by the controller, and then the feed-forward voltage is added to the comparator. Comparing with the triangle wave, the SPWM signal is obtained to control the on and off of the main circuit power tube. The method of generating SPWM by the device is to select the unipolar frequency SPWM mode, and improve the harmonic frequency of the SPWM waveform without increasing the switching frequency, so that the harmonic component of the output voltage can be obtained. Effective control.

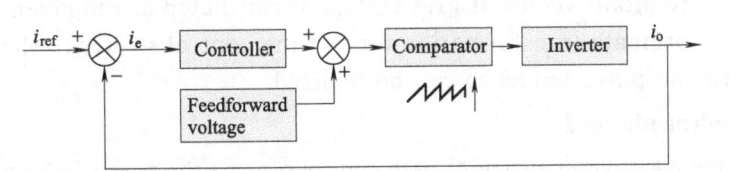

Figure 3. 7. 28   The principle diagram of current control strategy

The function of the voltage outer-loop is to track the direct current voltage command U_{dc} through the PI controller, and give the current inner loop instruction i_{ref}, where the direct current voltage instruction U_{dc} is a fixed constant, and the bus voltage U provided by the front-end Boost circuit is compared with it to generate the current given instruction i_{ref}.

(4) Other parameters affecting the quality of the current grid connection

The power supply quality of grid-connected inverter is not only related to the control strategy, but also affected by the current circulation when the grid-connected switch is closed, the dead zone time of inverter switch, the DC voltage, the power grid disturbance and the saturation and nonlinear characteristics of the isolation transformer core.

Before the grid-connected switch is closed, the output voltage of the inverter is not

85

consistent with the grid voltage including amplitude, phase, frequency, DC component and so on. Even if the given value of the incoming current is 0, there will be the current flowing between the inverter and the grid, which is called circulation here. Circulation includes DC circulation, fundamental component circulation, harmonic circulation, and the proportion of circulation is larger for light load, and smaller for heavy load. Duo to the fact that the equivalent output impedance and line impedance of the inverter are very small, the small voltage difference between the output voltage of the inverter and the grid voltage will bring a large circulation, and there will exist an inductance between the inverter and the power grid, which will make the circulation lag the voltage by 90 degrees, so the existence of this circulation will seriously affect the quality of current on line side. The inverter's control adopts PI regulation which can't realize non-static error track, so there is amplitude difference and phase difference between the inverter's output voltage and the grid voltage, therefore, the problem of reducing the circulation is transformed into the problem of reducing the static error of PI regulation.

In case where the frequency of inverter control switch is high, the effect of dead time on the output current waveform can not be ignored.

Voltage disturbance of power grid is a transient process, with continuous influence, while the non-linear saturation of the iron core often produces strong and continuous harmonic current disturbance in operation as the transformer is affected by such factors as manufacturing and operating conditions, which has great influence on the waveform quality of output current.

Parameter set and power quality analysis of grid-connected invert

Phase-lock precision: digital phase-lock method is used in this article. The sine form is made up of 4,000 points, and the phase lock precision is $360/4,000 = 0.09$, $\cos 0.09 = 0.99999$, so the phase-lock method can get excellent phase-lock effect, achieving higher *PF* value. The phase-lock is realized by tracking the resultant vector of grid voltage under rotating coordinates, that is to say, if the space position angle of the resultant vector of grid voltage is calculated as the given angle of the coordinate transformation used in the control strategy, the tracking of the power grid phase and phase sequence can be realized.

【Project Implementation】

① Turn on the main power switch of each control system; the power indicator light is shown.

② Put the brake of charger controller in the "RELEASE (auto brake)" state and close the "battery" air switch of "energy conversion storage and control system", and the battery is connected to PV MPPT control system while charging the charger controller. At this time, the charger controller is initialized, and the red indicator light is on (working in the braking state). The next operation can only be carried out after the red indicator light is off (exiting the braking state).

Practice of grid-connected inverter

③ Turn on the "frequency converter" switch of the "simulation energy control system" and power the frequency converter.

④ Press the "start-up" button (green button) of the control switch, operate the frequency converter to select the appropriate frequency and press RUN button to runs a wind energy simulator (frequency not greater than 20Hz).

⑤ Open "controller" switch of "grid-connected inverter control sys-

Section Ⅲ   Practical Training Project

tem", power on grid-connected inverter controller and initialize LCD.

⑥ Turn on the switch of "storage battery" and "grid-connected power generation", and the input and output voltage of inverter have numerical value.

⑦ Operate human-machine interface, select "parameter setting" by moving the cursor through the keyboard, namely, bus voltage $U = 120V$, current loop proportional coefficient $P = 1500$, current loop integral coefficient $I = 300$, feedforward voltage $FW = 38V$, dead time $DB = 2560\mu s$; click the "return" key, return to the initial interface, then click the "start/stop" button to start the inverter; record the values of each meter before and after the work of inverter and grid-connected;

S/N	Items	Inverter input		Inverter output	
		U/V	I/A	U/V	I/A
1	Before grid-connected				
2	After grid-connected				

⑧ After the inverter works normally, record the parameters of the inverter output electricity meter under no-load, DC load and AC load respectively:

S/N	Items	Inverter output power meter				
		U/V	I/A	P/kW	$Q/kV \cdot A$	PF
1	After grid-connected (no load)					
2	After grid-connected (motor)					
3	After grid-connected (LED light)					

⑨ Operate the human-machine interface, select "parameter setting" by moving the cursor through the keyboard, namely, bus voltage $U = 120V$, the current loop proportional coefficient $P = 150$, the current loop integral coefficient $I = 30$, the feed forward voltage $FW = 38V$, the dead time $DB = 2560\mu s$; click the "return" button, return to the initial interface, then click the "start/stop" button to start the inverter. By changing the current loop PID parameter setting, record the value of the inverter output meter in the table below, and draw the corresponding harmonic waveform:

S/N	Items	Inverter output power meter				
		U/V	I/A	PF	Voltage THD	Current THD
1	$P = 150$, $I = 30$					
2	$P = 500$, $I = 100$					
3	$P = 800$, $I = 180$					
4	$P = 1200$, $I = 250$					
5	$P = 1500$, $I = 300$					

⑩ Operate the human-machine interface, select "parameter setting" by moving the cursor through the keyboard, namely, bus voltage $U = 120V$, the current loop proportional coefficient $P = 1500$, the current loop integral coefficient $I = 300$, the feedforward voltage $FW = 38V$, the dead time $DB = 2560\mu s$; click the "return" button to return to the initial interface. By changing the setting of the dead-time DB parameter, the value of the output power table of the inverter is recorded in the following table:

S/N	Items	Inverter output power meter				
		U/V	I/A	PF	Voltage THD	Current THD
1	$DB = 2560\mu s$					
2	$DB = 2780\mu s$					
3	$DB = 2990\mu s$					
4	$DB = 3200\mu s$					

⑪ After ending the experiment, click the "start/stop" button to stop the inverter, and then operate the "STOP" key of the frequency converter to stop the fan.

⑫ Turn off the "stop" button (red button) and the "frequency converter" air switch for the control switch part of the "stimulation energy control system" in turn.

⑬ Turn off the "grid-connected power generation" "DC load" "AC load" "battery" "controller" air switch of "grid-connected inverter control system" and "battery" air switch of "energy conversion and storage control system" in turn, and then turn off the "main power" switch of each control system. If subsequent experiments are required, it is unnecessary to turn off the main power of the control system.

【Project Works】

① After the inverter works normally, change the U value of the bus voltage to observe the value of each meter.

② After the inverter works normally, observe the value of total harmonic distortion (THD) by changing the bus voltage value.

Project Ⅷ   Operation and Debugging of Photovoltaic Power Generation System

【Project Description】

Learn the key content of solar photovoltaic power generation technology, understand the basic principles of power generation of solar cell, and complete the operating and debugging of photovoltaic power generation system through the training platform.

【Ability goals】

① To be capable of grasping the principle of grid-connected photovoltaic power generation system.

② To be capable of grasping the role of solar panels, photovoltaic controllers, storage battery sets and grid-connected inverters played in grid-connected PV systems.

【Project Environment】

Photovoltaic power generation is a technique that converts optical energy directly into electrical energy using the photovoltaic effect of a semiconductor interface, and mainly consists of solar panels (components), controllers and inverters, its main components are electronic components. After the solar cells are connected in series, a large area of solar cell

Operation and debugging of photovoltaic power generation system

Section Ⅲ Practical Training Project

module can be formed by packaging protection, and then a photovoltaic power generation device can be formed with the power controller and other components.

(1) DC unit module

see project Ⅱ 【Project Environment】 (3), see P24.

(2) Sealed lead-acid battery

Technical indicators

Single battery capacity: 12V, 24Ah

Single battery size: 165mm×125mm×175mm

Physical map (Figure 3.8.1)

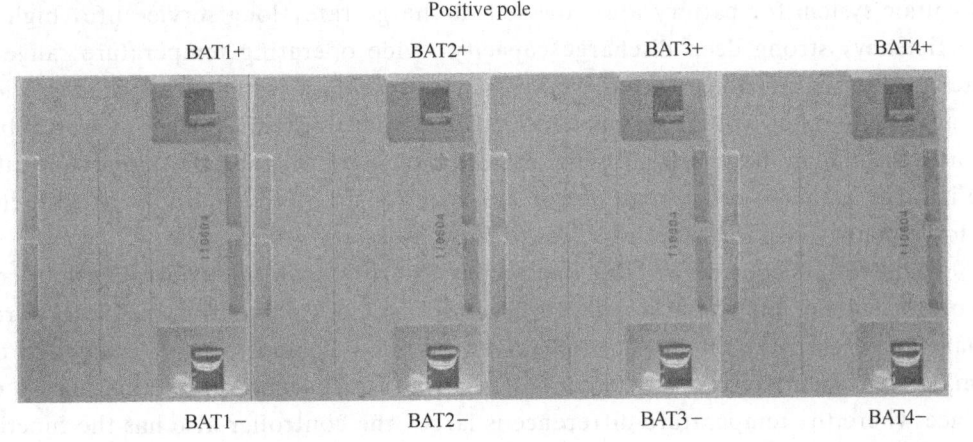

Figure 3.8.1 Storage battery

Port definition (Table 3.8.1)

Table 3.8.1 Port Definition of Storage Battery

S/N	Definition	Explanatory note
1	BATX＋	Positive output of battery
2	BATX－	Negative output of battery

(3) MPPT controller

see Project Ⅴ 【Project Environment】 (1), see P49.

【Project Principle and Basic Knowledge】

(1) Composition of solar photovoltaic power generation system

The power generation system which converts solar radiant energy into electricity through a solar cell is called a solar photovoltaic power generation system, or a solar cell power generation system. Although the application forms of solar photovoltaic power generation system are varied and the application scale is also wide from solar lawn lamp less than a watt and to a few hundred kilowatts or even a few megawatt of large-scale photovoltaic power station, but the solar photovoltaic power generation system's structure and working principle are basically the same. Its main structure consists of solar cell (or square array), storage battery (battery), photovoltaic controller, inverter (used when AC power is needed) and some ancillary facilities such as testing, monitoring and protection.

① Solar cell The solar cell, also known as the solar panel, is the core of the solar

89

power system and the most valuable part of the solar power system, whose function is to convert the radiant energy of sunlight into electrical energy and send it to the storage battery for storage or to be used to improve the working of the load directly. When the generating capacity is large, a series of multi-block battery is needed to form a solar cell matrix by series connection and parallel connection. The solar cells used at present are mainly crystalline silicon cells which are divided into monocrystalline silicon solar cells, polycrystalline silicon solar cells and amorphous silicon solar cells.

② Storage battery   The function of the storage battery is to store the electric energy from the solar cell and supply it to the load at any time. The basic requirements of solar photovoltaic system for battery are: low self-discharge rate, long service life, high charging efficiency, strong deep discharge capacity, wide operating temperature range, few maintenance or maintenance-free and low cost. At present, maintenance-free lead-acid batteries are used in photovoltaic systems. In small and micro systems, nickel-hydrogen batteries, nickel-cadmium batteries, lithium batteries or super capacitors are also available, while if large amounts of electric power are required to store, multiple battery clusters need to be connected in parallel or series to form a battery group.

③ Photovoltaic controller   The function of solar photovoltaic controller is to control the working state of the whole system. Its functions include: preventing the battery from overcharging, preventing the battery from over-discharging, short-circuit protection of the system, system polarity reverse protection, night anti-recharging protection and so on. In the place where the temperature difference is large, the controller also has the function of temperature compensation. In addition, the controller is also equipped with optical-controlled switch, time-controlled switch and other working mode as well as charging state, battery power and other various working state display functions. The photovoltaic controller is generally divided into small power, medium power, high power and wind-solar complementary controller and so on.

④ AC inverter   An AC inverter is a device that converts a DC from the output of a solar cell or battery into AC supply for the power grid or AC load. Inverter can be divided into independent operation inverter which is used for independent operation of solar power system and provides power supply for independent load and grid-connected inverter which is used in grid-connected solar power systems according to the mode of operation.

⑤ Ancillary facilities of photovoltaic power generation system   The ancillary facilities of photovoltaic power generation system include DC distribution system, AC distribution system, operation monitoring and testing system, lightning protection and grounding system etc.

(2) Working principle of solar photovoltaic power generation system

The solar photovoltaic power generation system can be divided into independent (off-grid) photovoltaic power generation system and grid-connected photovoltaic power generation system. Figure 3. 8. 2 shows a schematic diagram of the working principles of a stand-alone solar photovoltaic system. The core component of solar photovoltaic power generation is solar panels which convert sunlight into electricity directly and store it in storage batteries. When the load is used, the electric energy in the storage battery is reasonably distributed to each load through the controller. The current generated by solar cells is DC, which can be applied directly in the form of DC or converted into AC by AC inverters for the use of AC loads. The electricity generated by solar energy can be used instantly or stored in storage devices such as storage batteries and can be used when necessary.

Section Ⅲ　Practical Training Project

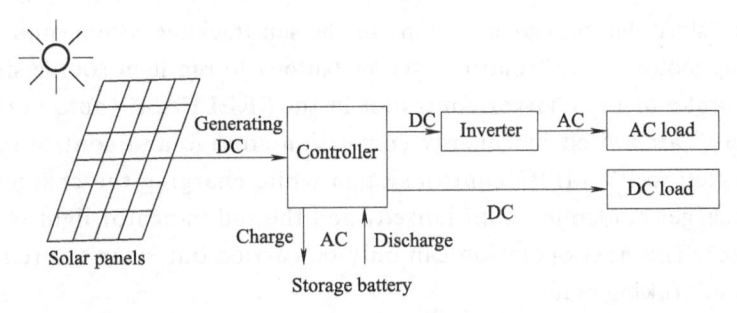

Figure 3. 8. 2　Working principle of stand-alone solar photovoltaic power generation system

Figure 3. 8. 3 shows a schematic diagram of the working principle of grid-connected solar photovoltaic systems. The grid-type photovoltaic power generation system can convert light energy into electric energy by the square array of solar cells, and the DC power distribution box into grid-connected inverter. The grid-connected inverter is composed of charge-discharge control, power regulation, AC inverter and grid-connected protection switching. The AC power output from the inverter is used for the load and the excess power is fed into the public power grid (known as the selling power) through such devices as power transformers. When the grid-connected PV system is short of electricity generation due to weather reasons or its own power consumption is too large, it can be supplied from the public grid to the AC load (called buy electricity). The system is also equipped with monitoring, testing and display system for the purpose of monitoring, testing of the whole system and recording data statistics such as power quantity, and can use computer network system to transmit control and display data remotely.

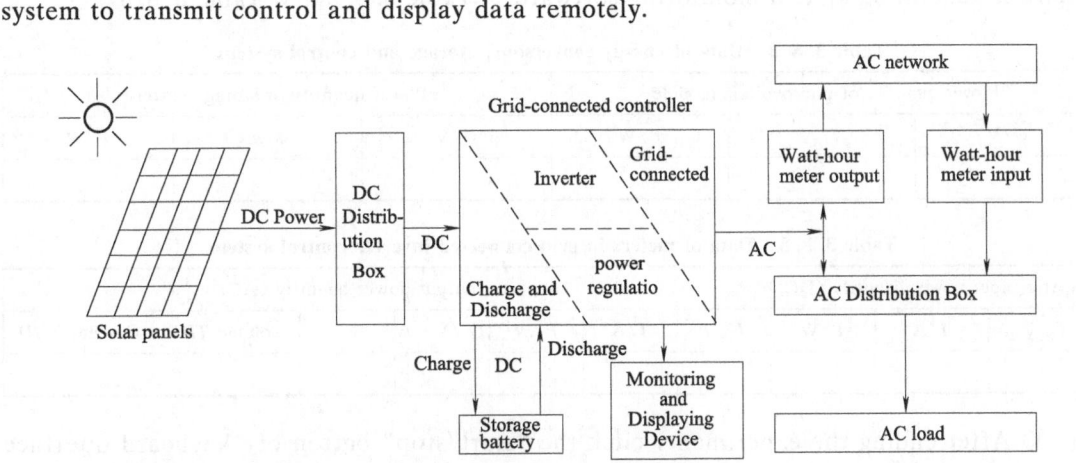

Figure 3. 8. 3　Working principle of grid-connected solar photovoltaic power generation system

【Project Implementation】

① Open the main power switch on each control cabinet, the system will be electrified and the three-phase power indicator light will be on.

② Open the "switching power supply" air switch of the "simulation energy control system" to make the switching power supply work.

③ Open the "simulation light source" air switch to make the sun lamp simulator open.

④ Open the "PLC" air switch and power the PLC, put the slide switch on the PLC into the "RUN" state to make the PLC program run.

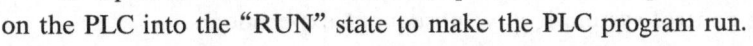

Running and debugging
of photovoltaic
generation system

91

⑤ Press the "start" button (green button) of the sun tracking system control switch to prepare each stepping motor, press "control" (yellow button) to run light source simulator.

⑥ Put the brake of the charger controller in the "RELEASE (auto brake)" state and close the "battery" air switch in "energy conversion storage and control system" and the battery is connected to PV MPPT control system while charging the charger controller. At this time, the charger controller is initialized, and the red indicator light is on (working in the braking state). The next operation can only be carried out after the red indicator light is off (exiting the braking state).

⑦ Open the "PV output" and "PV MPPT" air switch and press the reset button K1 on the "CPU core module" to make the system reset.

⑧ Press the "DOWN" button of the "Human-Computer Interaction Module" to switch "Power Tracking CVT" and then click "ENTER" button, and then the system will automatically perform power tracking.

⑨ Open the "control unit" air switch of "grid-connected inverter control system" and power on the grid-connected inverter to initialize LCD.

⑩ Open the air switch of "storage battery" and "grid-connected power generation" to display the value of each intelligent meter.

⑪ Press the "start/stop" button of the "keyboard interface module", the grid-connected inverter begins to work, and the working light is on. At this time, air switch the "direct current load" and the "AC load" can be opened to observe the power change of inverter output.

⑫ Record the values of the smart meters in the "THNRFG-4 wind-solar complementary power generation system monitoring software" (Table 3.8.2 and Table 3.8.3).

Table 3.8.2　Data of energy conversion, storage and control system

Power quantity of photovoltaic module			Power quantity of storage battery		
U/V	I/A	P/W	U/V	I/A	P/W

Table 3.8.3　Data of meters in grid-connected inverter control system

Inverter input power quantity (DC)			Inverter output power quantity (AC)						
U/V	I/A	P/W	U/V	I/A	P/W	$Q/V \cdot A$	PF	Voltage THD	Current THD

⑬ After ending the experiment, click the "start/stop" button of "keyboard interface module" in "grid-connected inverter control system" to make the inverter stop work, and then click the "stop" button (red button) of sun tracking system control switch of the "stimulation energy control system" to stop the sun tracking system.

⑭ Turn off the "light source stimulator", "switching power supply" and "PLC" air switch of "stimulation energy control system" in turn.

⑮ Turn off the air switch of "PV output" "PV MPPT" and "Battery" of "energy conversion storage and control system" in turn.

⑯ Shut down the "grid-connected power generation" "DC load" "AC load" "battery" and "controller air switch" in "grid-connected inverter control system" in turn, and finally shut down the "main switch" button of all control systems. It is not necessary to turn off the main power supply of the control system if any subsequent experiments.

Section Ⅲ　Practical Training Project

【Project Works】

Draw the working principle flow diagram of the solar photovoltaic power generation system in the experimental platform and write the process of energy transfer.

Project Ⅸ　Operation and Debugging of Wind -Solar Complementary Generating System

【Project Description】

Learn the key speed of wind-solar complementary control, enable to independently set the parameters of the grid connected inverter of the system and analyze the power quality.

【Ability goals】

① To be capable of grasping the principle of wind-solar complementary power generation system.

② To be capable of grasping the function of each component of wind-solar complementary power generation system.

【Project Environment】

Operation and debugging of wind-solar complementary generating system

The operation and the debugging of wind-solar complementary power generation system is composed of a simulation energy control system, an energy conversion storage and control system, a grid-connected inverter control system and an energy monitoring management system.

① The stimulation energy control system (Table 3. 9. 1) consists of a control panel (composed of power supply, mesh board, tool drawer), a programmable logic controller (PLC), programming lines, an stimulation module, a frequency converter, a touch screen, a AC contactor, a relay, a button, a switch and so on.

Table 3. 9. 1　Stimulation energy control system

S/N	Name	Model and specification	Quantity	Remark
1	Control panel	800mm×600mm×1880mm	1 set	
2	PLC	FX3U-32MT transistor host Input：16 points，Output：16 points	1	Mitsubishi
3	Stimulation module	FX3U-3A-ADP Input channel：2，Output channel：1 Maximum resolution：12 bits Range：DC0～5V/10V，DC4～20mA	1	Mitsubishi
4	Frequency converter	FR-A800 Rated power：0.75kW Rated input voltage：3-phase AC380～480V 50/60Hz Rated output voltage：3-phase AC380～480V Output frequency：0.2～400Hz	1	Mitsubishi

93

② The energy conversion storage and control system (Table 3.9.2) consists of a control panel (power supply, mesh board, tool drawer), a photovoltaic array bus module, a DC lightning arrester, a DC voltage intelligent digital display meter, a DC current intelligent digital display, a disk resistor, a DC voltage and current sampling module, a CPU core module, a human-computer interaction module, a PWM driving module, a communication module, a wireless communication module, a temperature alarm module, DC-DC Boost/Buck/Boost-Buck three main circuit modules, the storage battery, a charger controller, a 51 ISP downloader, a PIC programmer and so on.

Table 3.9.2　Energy conversion storage and control system

S/N	Name	Model and specification	Quantity	Remark
1	Control panel	800mm × 600mm × 1880mm	1 set	
2	DC voltage intelligent digital display meter	Input range: 0～600V; Accuracy: 0.5 level	2 pcs	RS-485 interface
3	DC current intelligent digital display meter	Input range: 0～5A Accuracy: 0.5 level	2 pcs	RS-485 interface
4	Photovoltaic array bus module	4-circuit	1	
5	DC lightning arrester	Working voltage: 24V DC Discharge current: 5kA	1	
6	Disk resistor	$2 \times 72\Omega, 1.45A$	1	
7	CPU core module	CPU chip: AT89S52 (supporting ISP download) stimulation input channel: 8-circuit Switch volume: 8-circuit Serial communication interface: 1-circuit Human-machine interface: 1-circuit	1	
8	Human-computer interaction module	Independent button: 5 12864 character LCD	1	
9	PWM driving module	CPU chip: PIC16F690 Serial communication port: 1-circuit Switch volume: 8-circuit Isolated PWM driving signal: 2-circuit	1	
10	Communication module	RS-232 interface: 2-circuit RS-485 interface: 2-circuit	1	
11	Wireless communication module	ZigBee frequency range: 2405～2480MHz transmission distance: >10m Bluetooth frequency range: 2.402～2.480GHz; it can support multiple Bluetooth devices at the same time WiFi supports 802.11b/g/n wireless standard supports two data transfer modes of transparent/protocol, supports heartbeat signal and shows connection indication	2pcs	
12	Temperature alarm module	Optocoupler isolated relay output: 3-circuit Temperature sensor: 1-circuit	1	
13	DC voltage and current sampling module	Voltage channel: 1-circuit Current channel: 1-circuit	2 pcs	

Continued

S/N	Name	Model and specification	Quantity	Remark
14	DC-DC main circuit module	Boost main circuit	1	
		Buck main circuit	1	
		Boost-Buck main circuit	1	
15	Storage battery	12V/24Ah sealed lead-acid battery	4 pcs	
16	Charger controller	Rated voltage: 12V, 24V automatic identification Maximum load current: 35. 61A/17. 80A Cut-off voltage for fully charged battery: 0~15V/30V can be set	1	
17	51 ISP downloader	Support 89S51/52 online download	1 set	
18	PIC programmer	Support PIC series chip programming	1 set	

③ The grid-connected inverter control system（Table 3.9.3）consists of a DSP core module, an interface module, a LCD module, a keyboard interface module, a drive circuit module, a Boost circuit module, a bus voltage sampling module, a grid voltage sampling module, a current sampling module, a DC load, an AC load, a DC voltage intelligent digital display meter, a DC current intelligent digital display meter, an inverter output power meter, an isolation transformer, an off-grid inverter, a DSP simulator and so on.

The grid-connected inverter converts DC24V into AC36V and 50Hz, and increases into AC220V connected with the single-phase electric supply through the transformer. The main controller adopts TMS320F2812 chip of TI fixed point 32-bit, and the output power factor is close to 1. Double closed-loop control is adopted, in which the inner loop is the current loop and the outer loop is the voltage loop. The digital phase-lock technology is adopted to synchronize the network.

Table 3.9.3 Parallel inverter control system

S/N	Name	Model and specification	Quantity	Remark
1	Control Panel	800mm×600mm×1880mm	1 set	
2	DC Voltage Intelligent Digital Display Meter	Input range: 0~600V; Accuracy: 0. 5 level	1 piece	RS-485 interface
3	DC Current Intelligent Digital Display Meter	Input range: 0~5A; Accuracy: 0. 5 level	1 piece	RS-485 interface
4	Inverter Output Meter	Input network: single-phase 2-circuit Input frequency: 45~65Hz; Input voltage rating: AC100V; Input current rating: AC5A Measurement accuracy: Frequency 0. 05Hz, reactive power 1, other 0. 5	1 piece	RS-485 interface
5	DC Load	Rated voltage: DC24V	1 piece	
6	AC Load	Rated voltage AC/DC 36V	1 piece	
7	Isolation Transformer	Input voltage 36V, Output voltage 220V Frequency: 50Hz	1 piece	
8	Drive Circuit Module	IPM intelligent power module: 6MBP20RH060 Rated current: 20A DC bus voltage: 450V Maximum switching frequency: 20kHz	1 piece	

Continued

S/N	Name	Model and specification	Quantity	Remark
9	Boost Circuit Module	DC boost	1 piece	
10	DSP Core Module	CPU: TMS320LF2812 Main frequency: 150MHz RAM: 4Mb (256K×16Bit) Flash: 8Mb(512K×16Bit)	1 piece	
11	LCD Module	240×128 LCD screen	1 piece	
12	Interface Module	Analog input: 4-circuit Serial communication interface: 1-circuit Keyboard interface: 1-circuit LCD interface: 1-circuit	1 piece	
13	Keyboard Interface Module	4×4 matrix keyboard	1 piece	
14	Bus Voltage Sampling Module	Voltage sampling channel: 1-circuit	1 piece	
15	Grid Voltage Sampling Module	Voltage sampling channel: 1-circuit	1 piece	
16	Current Sampling Module	Current sampling channel: 1-circuit	1 piece	
17	Off-grid Inverter	Input voltage: DC24V Efficiency: >85% Output voltage: AC220V Rated power: 300W Output waveform: sine wave Output frequency: 50Hz Waveform distortion: 3%	1 piece	
18	DSP Simulator	Support DSP on-line simulation and download	1 set	

④ The energy monitoring and management system （Table 3. 9. 4） consists of a system controller core module, a relay module, a communication module, a 15-inch industrial tablet computer, a configuration software and so on, and can communicate with each control system. The host computer software can display the running data in real time, and can change the running state automatically or manually according to the control requirement.

Table 3. 9. 4　Energy monitoring and management system

S/N	Name	Model and specification	Quantity	Remark
1	Control Panel	800mm×600mm×1880mm	1 set	
2	15-inch Industry Tablet Computer	Display: 15-inch liquid crystal CPU: Intel Atom N270, 1. 6 GHz Motherboard: Microstar A9830IMS Memory: DDR2 1G Hard disk: 160G Touch screen: 4-circuit resistive touch screen PC card: 10~100M Serial port: 5 USB: 4 VGA interface: 1 PCI slot: 1	1	

Section III   Practical Training Project

Continued

S/N	Name	Model and specification	Quantity	Remark
3	System Controller Core Module	Analog Input: 3-circuit Stimulation resolution: 8bits Switching output: 5-circuit Support ISP download Serial communication interface: 1-circuit Battery quantity monitoring indicator Support battery charge simulation	1	
4	Relay Module	Optocoupler isolate relay output: 5-circuit	1	
5	Configuration Software	Industrial configuration software	1 set	

【Project Principle and Basic Knowledge】

Wind power and solar power are both affected by the weather environment with certain limitations, but there is a certain degree of complementarity between them. The so-called "wind-solar complementary" refers to the alternate use of solar and wind power during the day and at night. Generally speaking, in sunny days, the wind may be relatively small, so it is mainly solar power generation and is supplemented by wind power generation. At night, it is powered only by wind turbines and has greater winds than that during the day. As a result, the complementary forms of power generation are formed throughout the day. The best portfolio will be obtained by using a proper proportion of solar and wind power, and the specific proportion of the combination should be comprehensively demonstrated through the technology, investment, meteorology, and load situation. Although "wind-solar complementary" power generation can form a certain complementary relationship, it is still affected by meteorological conditions, and the stability can be improved significantly if the storage battery is added. However, the battery belongs to the energy storage element. Although energy storage, as energy storage element, can make up the instability of the two power supply by using it to a certain extent, it has limit stored energy and can't supply power continuously for a long time, in case of continuous rainy and windless weather, the power supply capacity of the whole power supply system will be greatly reduced. For loads that are important or require higher stability of power supply, it is necessary to consider the use of spare diesel generator sets to form an integrated power supply system of fans, photovoltaic and diesel generators, thus enhancing the reliability and stability of power supply.

【Project Implementation】

① Open the main power switch on each control cabinet, the system will be electrified and the three-phase power indicator light will be on.

② Open the air switch of "stimulate power supply", "light source stimulator" and "frequency converter" of "stimulation energy control system".

③ Open the "PLC" air switch and power the PLC, put the slide switch on the PLC into the "RUN" state to make the PLC program run.

④ Press the "start" button (green button) of the sun tracking system control switch to prepare each stepping motor, press "control" (yellow button) to run light source simulator.

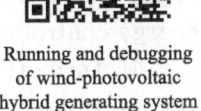

Running and debugging of wind-photovoltaic hybrid generating system

97

⑤ Put the brake of the charger controller in the "RELEASE (auto brake)" state and close the "battery" air switch in "energy conversion storage and control system" and the battery is connected to PV MPPT control system while charging the charger controller. At this time, the charger controller is initialized, and the red indicator light is on (working in the braking state). The next operation can only be carried out after the red indicator light is off (exiting the braking state).

⑥ Open the "PV output" and "PV MPPT" air switch and press the reset button K1 on the "CPU core module" to make the system reset.

⑦ Press the "DOWN" button of the "Human-Computer Interaction Module" to switch "Power Tracking CVT" and then click "ENTER" button, and then the system will automatically perform power tracking.

⑧ Press the green start button on the "stimulation energy control system" to power the frequency converter (Note: When the red indicator light of the charger controller is on, the frequency converter cannot be started, because the charger controller works in the braking state). Press PU/EXT button, it enters PU operation mode. Set the frequency under 20Hz through the M knob and press RUN key to run.

⑨ Open the "controller" air switch of "grid-connected inverter control system" and power on the grid-connected inverter to initialize LCD.

⑩ Open the air switch of "storage battery" and "grid-connected power generation" to display the value of each intelligent meter.

⑪ Record the values of the smart meters in the "THNRFG-4 wind-solar complementary power generation system monitoring software" (Table 3. 9. 5 and Table 3. 9. 6).

Table 3. 9. 5 Data of energy conversion, storage and control system

Power quantity of photovoltaic module			Power quantity of storage battery		
U/V	I/A	P/W	U/V	I/A	P/W

Table 3. 9. 6 Data of meters in grid-connected inverter control system

Inverter input power quantity (DC)			Inverter output power quantity (AC)						
U/V	I/A	P/W	U/V	I/A	P/W	Q/VA	PF	Voltage THD	Current THD

⑫ At the end of the experiment, click the "start/stop" button of the "keyboard interface module" of the "grid-connected inverter control system" to make the inverter stop working, then operate the "STOP" button of the frequency converter to stop the fan running, and finally click the "stop" button (red button) of the "sun-tracking control switch" of the "stimulation energy control system" to stop the sun- tracking system.

⑬ Operate the stop button (red button) of control switch of for "stimulation energy control system" and disconnect the inverter power; turn off the "frequency converter", "light source simulator", "switching power supply" and "PLC" air switch of "stimulation energy control system" in turn.

⑭ Turn off the air switch of "PV output", "PV MPPT" and "battery" of "energy conversion storage and control system" in turn.

Section III   Practical Training Project

⑮ Turn off the "grid-connected power generation", "battery" and "controller" air switch of the "grid-connected inverter control system" in turn, and finally shut down the "main switch" button of all control systems. It is not necessary to turn off the main power supply of the control system if any subsequent experiments.

【Project Works】

① Draw the working principle flow diagram of the wind-solar complementary power generation system of the experimental platform, and write the process of energy transfer.

② Describe the composition and advantages and disadvantages of photovoltaic and wind-solar complementary power generation systems.

Project X   Configuration Design of Energy Monitoring and Management System

【Project Description】

It mainly completes the work of real-time database configuration, acquisition system configuration, data service configuration, screen configuration, network site configuration, report development, script programming and common functional parameters configuration.

【Ability goals】

① To be capable of using the configuration software of Symlink.

② To be able to independently complete secondary development of software.

【Project Environment】

(1) Energy monitoring and management unit

The energy monitoring and management unit mainly aims at the monitoring and controlling of the whole process, such as the battery power and the wind-solar complementary system working state etc.

Energy monitoring and management

(2) Composition of energy monitoring and management unit

The functional flow diagram of energy monitoring and management unit is shown in Figure 3.10.1. It consists of a system controller module, a relay module, a communication interface module and a wind-solar complementary power system monitoring software. The system controller module can display the battery power in real-time, carry on the working state shift of wind-solar complementary system, can switch the input to the battery power simulator, and can switch the working state of the wind solar complementary system by giving the battery power, so as to avoid waiting for the battery charging and discharging for a long time. The monitoring software of wind-solar complementary power generation system completes the monitoring and control of the whole system (extended functions).

System controller module: it mainly includes a single-chip microcomputer, a battery power detection module, a battery power simulator, a power value display and other components.

Relay module: it mainly includes a photocoupler isolate, a relay and a LED indicator light.

Communication interface module: completes protocol conversion of RS-232/485 com-

99

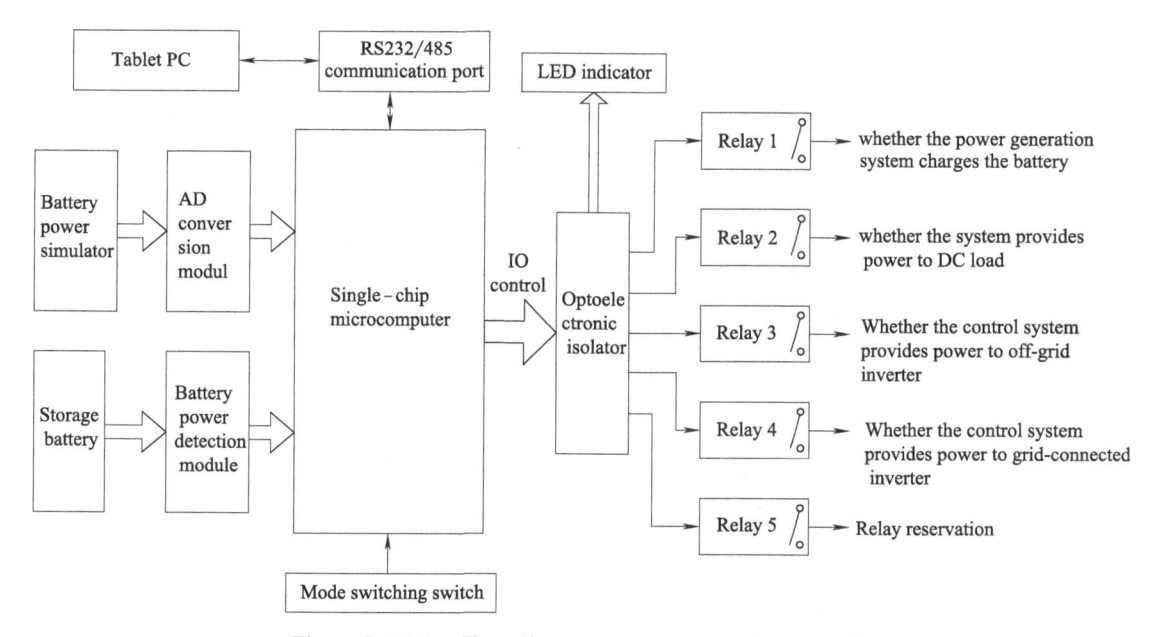

Figure 3.10.1 Flow diagram of energy monitoring unit

munication interface (see extended functions in Project 2 for deatil).

Wind-solar complementary power system monitoring software: completes the whole system monitoring and control and displays the data of each power meter.

(3) System controller module

The structure flow diagram of the system controller is shown in Figure 3.10.2. It mainly consists of the main controller, the battery power detection module, the battery power simulator, the battery power display and the output control contact.

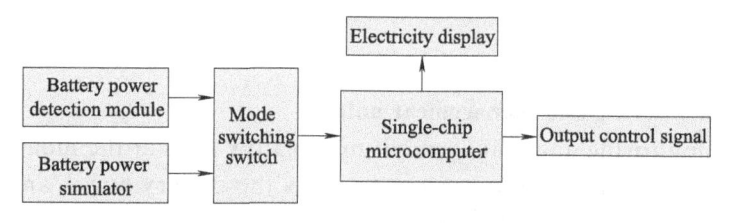

Figure 3.10.2 Flow diagram of system controller structure

The modules are described as follows.

① Main Controller: it is composed of a single-chip microcomputer;

② Battery Power Simulator: it is composed of adjustable potentiometer and AD conversion unit to realize the simulation of battery power;

③ Power Display Terminal: it consists of a light-emitting diode (LED) and a latch, displaying the quantity of electricity detected;

④ Output Control Signal: the control signal output by single chip microcomputer according to battery power;

⑤ Battery Power On-line Detection Module: the detection of the battery power;

⑥ Switch SW1: When the switch is turned to "0", the battery power simulator is selected; when the switch is turned to "1", the battery power online detection module is selected.

The schematic diagram and port definitions are shown in Figure 3.10.3, Figure 3.10.4 and Table 3.10.1.

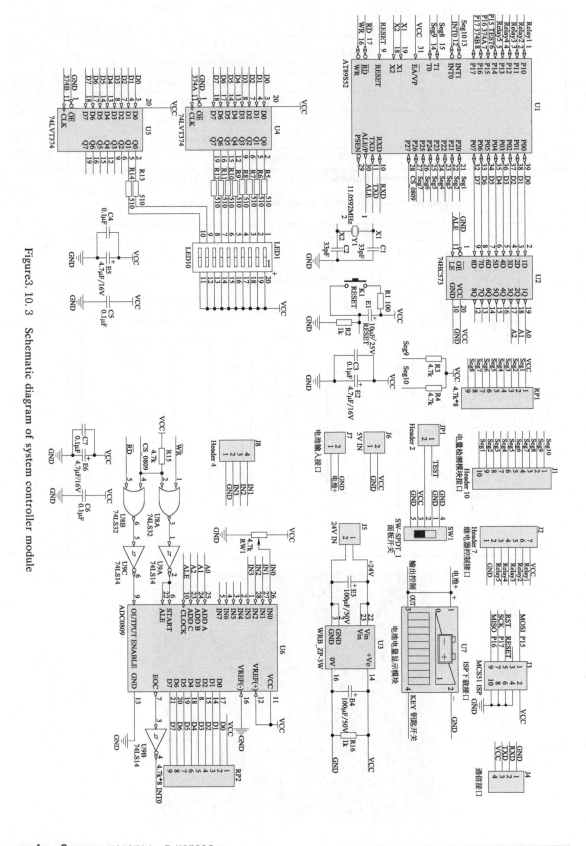

Figure3.10.3 Schematic diagram of system controller module

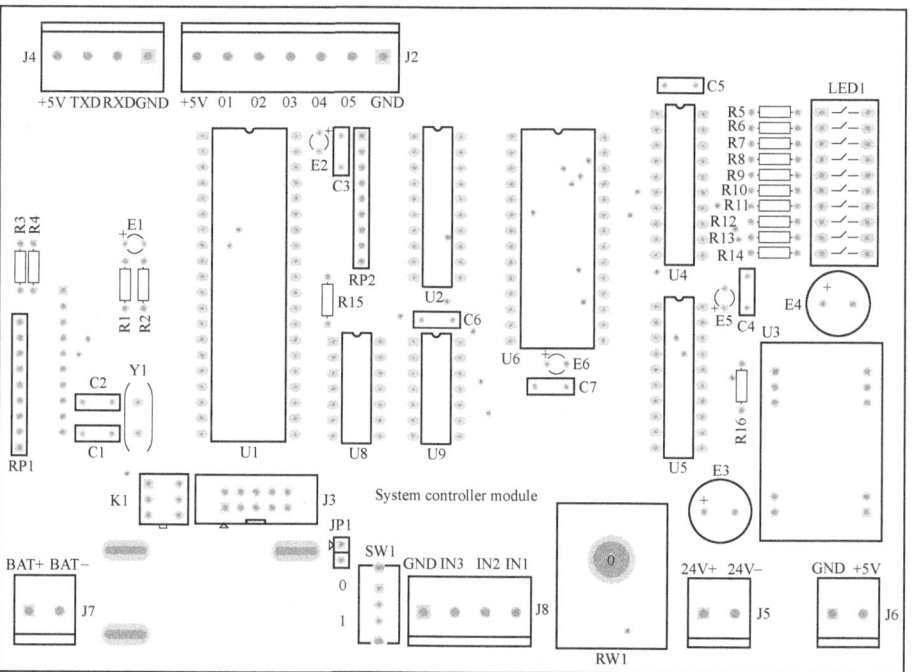

Figure 3.10.4　Physical diagram of system controller module

Table 3.10.1　Definition of system controller module port

S/N	Definition	Explanatory Note	Extended Interface	Remark
1	J2: +5V	+5V power output		
2	J2: O1	Switch output interface 1		
3	J2: O2	Switch output interface 2		
4	J2: O3	Switch output interface 3		
5	J2: O4	Switch output interface 4		
6	J2: O5	Switch output interface 5		
7	J2: GND	Ground		
8	J3	89S52 online download interface		

Continued

S/N	Definition	Explanatory Note	Extended Interface	Remark
9	J4: +5V	+5V power output		
10	J4: TXD	Transmitting terminal of serial port		
11	J4: RXD	Receiving terminal of serial port		
12	J4: GND	Ground		
13	J5: 24V+	Isolated power supply DC/DC inputs positive 24V		
14	J5: 24V-	Isolated power supply DC/DC inputs negative 24V		
15	J6: +5V	+5V power output	√	
16	J6: GND	Ground	√	
17	J7: BAT+	Battery (Positive)		
18	J7: BAT-	Battery (Negative)		
19	J8: IN1	Analog input AD0809 channel 1	√	
20	J8: IN2	Analog input AD0809 channel 2	√	
21	J8: IN3	Analog input AD0809 channel 3	√	
22	J8: GND	Ground	√	
23	JP1	Operation procedure: short circuit: download program: open circuit		

(4) Relay module

The relay module consists of a isolation device, a LED, and a power relay. The LED indicates the state of each output control point (that is, the relay).

The schematic diagram and port definitions are shown in Figure 3.10.5, Figure 3.10.6 and Table 3.10.2.

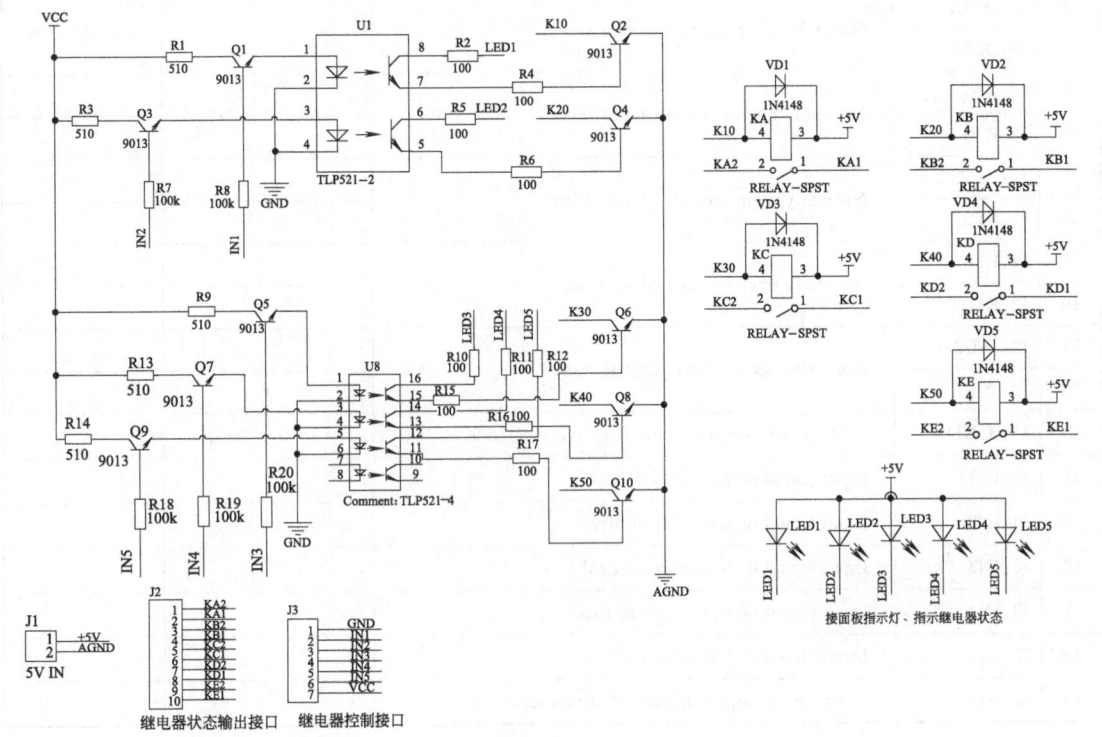

Figure 3.10.5　Schematic diagram of relay module

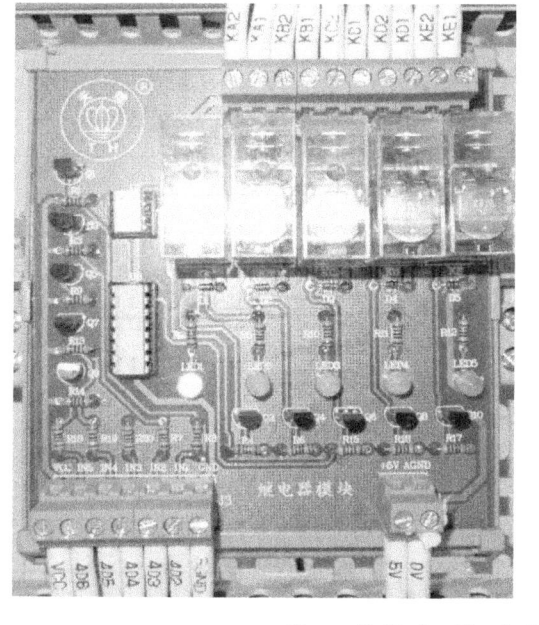 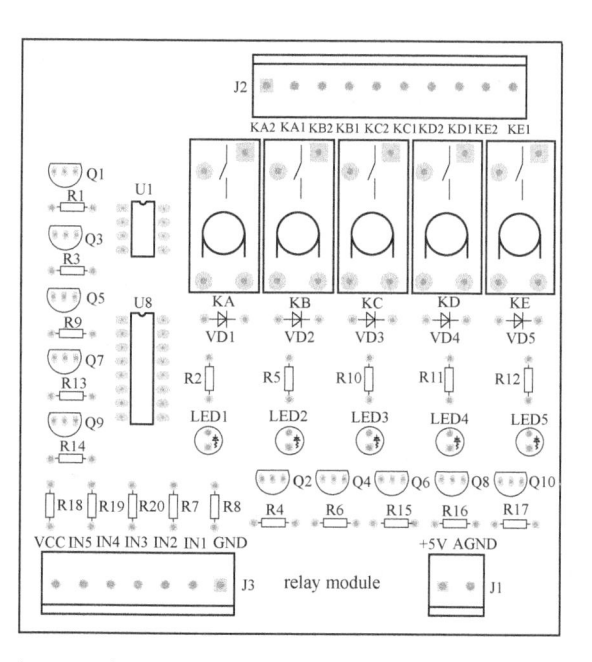

Figure 3.10.6   Physical photograph of relay module

Table 3.10.2   Port definition of relay module

S/N	Definition	Explanatory Note	Extended Interface	Remark
1	J1: +5V	Isolation +5V power supply inputs positive electrode		
2	J1: AGND	Isolation +5V power supply inputs negative electrode		
3	J2: KA2	Normally open contact of KA relay		
4	J2: KA1			
5	J2: KB2	Normally open contact of KB relay		
6	J2: KB1			
7	J2: KC2	Normally open contact of KC relay		
8	J2: KC1			
9	J2: KD2	Normally open contact of KD relay		
10	J2: KD1			
11	J2: KE2	Normally open contact of KE relay		
12	J2: KE1			
13	J3: GND	+5V power supply inputs negative electrode		
14	J3: IN1	Input signal of KA relay control		
15	J3: IN2	Input signal of KB relay control		
16	J3: IN3	Input signal of KC relay control		
17	J3: IN4	Input signal of KD relay control		
18	J3: IN5	Input signal of KE relay control		
19	J3: VCC	+5V power supply inputs positive electrode		

Section III   Practical Training Project

Figure 3. 10. 7   Physical diagram of tablet PC

(5) Tablet PC

The physical picture of tablet PC is shown in Figure 3. 10. 7.

The technical specifications tablet PC are as follows:

① Case Material: 8MM hard oxidized aluminum alloy surface;

② Processor: Intel celerom M 900MHz/2M Cache;

③ Chip set: Intel 82852 Bridge/Intel Ich 4 South Bridge;

④ I/O Chip: Winbone W83627DHG-A;

⑤ System Memory: standard 512M (DDR 200/266) memory;

⑥ BIOS: Award 4 Mb Flash ROM BIOS;

⑦ Expansion: 1 PCI slot (PCI is compatible with Rev. 2. 2), and supporting 3 main PCIs;

⑧ USB: 4 USB 2. 0 interface;

⑨ DIO 1: 6-bit GPIO for DI and DO;

⑩ Monitor: industrial-level LCD with 1024×768 resolution;

⑪ Product Size: 280mm×210mm×60mm ($D \times W \times H$);

⑫ MIO: 1×VGA、1×S VIDEO-OUT、1×RJ45、4×RS-232、1×AUDIO-OUT、1×CF Set、1×PS/2   KEY BOARD/MOUSE、1×4P-DC;

⑬ Power Supply: DC12V input.

(6) Interface converter

CONV485 is a cost-effective interface converter for RS-232 signal and RS-485 signal, and is fed directly from the device's serial port (such as the COM port of the computer) without external power supply. Interface drawing and physical drawing are shown in Figure 3. 10. 8.

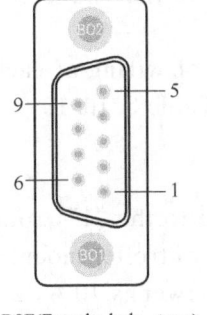
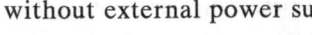

DB9F(Female,hole–type)   DB9M(male,needle)

Figure 3. 10. 8   Interface drawing and physical drawing of CONV485

The technical indicators

Interface Standard: EIA RS-232, RS-485;

RS-232 Signal: TXD, RXD, GND;

RS-485 Signal: D −, D +, GND;

Working Mode: point-to-point, two-line half duplex of point-to- multi point;

Transmission Distance: 5 meters for RS-232 Port, 1,200 meters for RS-485 Port;

Direction Control: adopt data flow automatic control (DSAC);

Load Capability: support 32points;

Interface Protection: 600W surge protection, 15kV ESD protection;

Interface Form: RS-232 for DB9F, RS-485 for DB9M.

The RS-485 interface adopts DB9M and the RS-232 port adopts DB9F interface, as shown in Table 3.10.3.

Table 3.10.3 Port definition of interface converter

	Pin Number	Explanatory Note
RS-232 Port (DB9F)	2	RXD
	3	TXD
	5	GND
RS-485 Port (DB9M)	1	485 −
	2	485 +
	5	GND

【Project Principle and Basic Knowledge】

The energy monitoring software configures the status data of the hardware (controller module, relay module and interface conversion module) of the management unit, including the configuration of the acquisition system, data service, screen, network station, reporting development, script programming and the configuration of common function parameters.

【Project Implementation】

(1) Installation and operation of energy monitoring and management unit

The energy monitoring and management unit includes the controller module, the relay module and the interface conversion module. The installation and operation procedures are as follows:

① Installation   According to the following layout diagram (Figure 3.10.9), the aluminum lead rail, the slot, the function modules, the terminals, the connection modules and so on are fixed to the mesh board for preparing the wiring.

② Wiring   The definition of terminal block and the table of wiring of each parts are shown in Figure 3.10.10, Figure 3.10.11 and Table 3.10.4~Table 3.10.10.

③ Operational instructions

Work state control (extended function)

Turn the "SW1" dip-switch of the system controller module to the "1", rotate "RW1" potentiometer and observe the LED light column of the system controller module (the light column from the bottom to the top indicates that the battery power is 10%, 20%, 90%, 100%) and the change of LED of relay module shall meet the following requirements:

• When the battery power display value is <50%, the relay KA is closed, and the LED 1 is on.

• When the battery power display value is ≥50% and <70%, the relay KA and KB are closed, and the LED1 and LED2 are on.

• When the battery power display value is ≥70% and <90%, the relay KA, KB and KC are closed, and the LED1, LED2 and LED3 are on.

Section Ⅲ　Practical Training Project

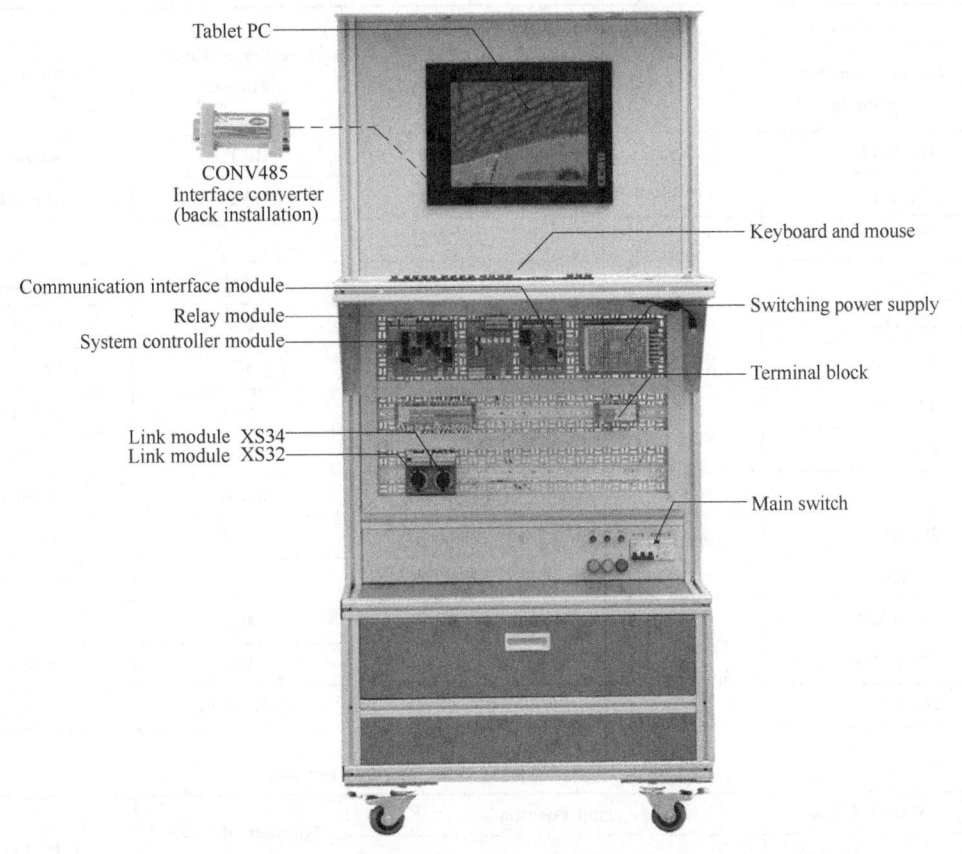

Figure 3. 10. 9　Layout of energy monitoring unit components

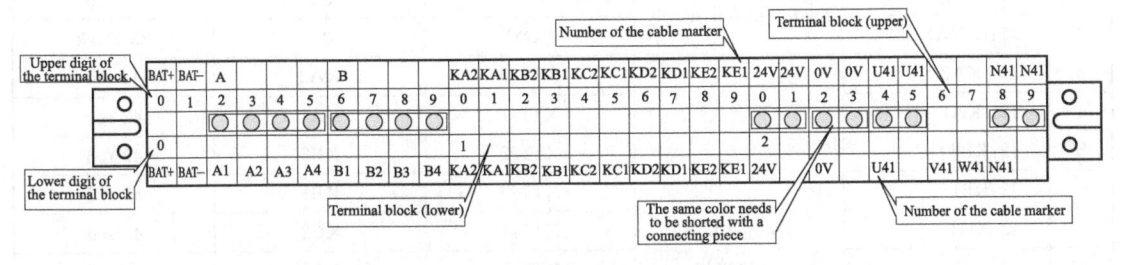

Figure 3. 10. 10　Port definition of terminal block

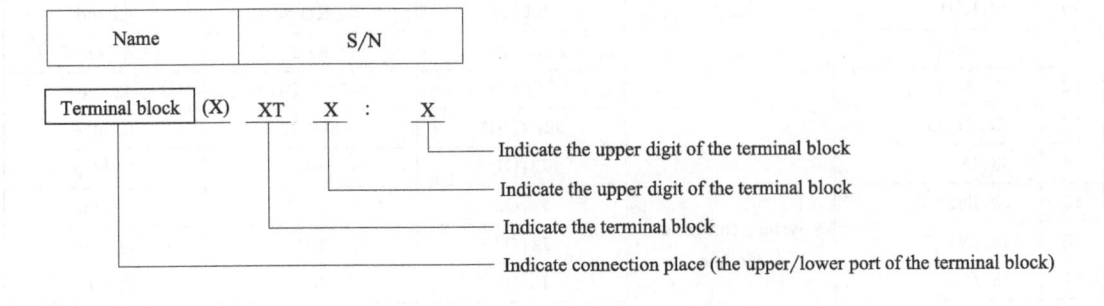

Figure 3. 10. 11　Number definition of terminal block

107

Table 3.10.4　Wiring list of system controller module

S/N	Start Position System Controller Module	End Position Name	End Position Number	Number of Cable Marker	Line Type
1	J7: BAT+	Terminal block (upper)	XT0:0	BAT+	42 red
2	J7: BAT-		XT0:1	BAT-	42 black
3	J2: +5V	Relay module	J3: VCC	VCC	12 blue
4	J2: O1		J3: IN1	402	12 blue
5	J2: O2		J3: IN2	403	12 blue
6	J2: O3		J3: IN3	404	12 blue
7	J2: O4		J3: IN4	405	12 blue
8	J2: O5		J3: IN5	406	12 blue
9	J2: GND		J3: GND	SGND	12 blue
10	J5: 24V+	Terminal block (upper)	XT2:0	24V	23 red
11	J5: 24V-		XT2:2	0V	23 black
12	J4: TXD	Communication interface module	J9: T1IN	400	12 blue
13	J4: RXD		J9: R1OUT	401	12 blue
14	J4: GND		J9: GND	GND (GND)	12 blue

Table 3.10.5　Wiring table of relay module

S/N	Start Position Relay Module	End Position Name	End Position Number	Number of Cable Marker	Line Type
1	J1: +5V	Switch power supply	+5V	5V	23 red
2	J1: AGND		COM	0V	23 black
3	J2: KA2	Terminal block (upper)	XT1:0	KA2	42 red
4	J2: KA1		XT1:1	KA1	42 red
5	J2: KB2		XT1:2	KB2	42 red
6	J2: KB1		XT1:3	KB1	42 red
7	J2: KC2		XT1:4	KC2	42 red
8	J2: KC1		XT1:5	KC1	42 red
9	J2: KD2		XT1:6	KD2	42 red
10	J2: KD1		XT1:7	KD1	42 red
11	J2: KE2		XT1:8	KE2	42 red
12	J2: KE1		XT1:9	KE1 (KE1)	42 red
13	J3: GND	System controller module	J2: GND	SGND	12 blue
14	J3: IN1		J2: O1	402	12 blue
15	J3: IN2		J2: O2	403	12 blue
16	J3: IN3		J2: O3	404	12 blue
17	J3: IN4		J2: O4	405	12 blue
18	J3: IN5		J2: O5	406	12 blue
19	J3: VCC		J2: +5V	VCC	12 blue

Section Ⅲ   Practical Training Project

Table 3.10.6   Wiring table of communication interface module

S/N	Start Position	End Position		Number of Cable Marker	Line Type
	Relay Module	Name	Number		
1	J1:24V+	Terminal block (upper)	XT2:1	24V	23 red
2	J1:24V−		XT2:3	0V	23 black
3	J9:T1IN	System controller module	J4:TXD	400	12 blue
4	J9:R1OUT		J4:RXD	401	12 blue
5	J9: GND		J4: GND	GND	12 blue

Table 3.10.7   Wiring table of switching power supply

S/N	Initial Position	End Position		Number of Cable Marker	Line Type
	Switch Power Supply	Name	Number		
1	L	Terminal block (upper)	XT2:4	U41	23 red
2	N		XT2:8	N41	23 black
3	EARTH	Control cabinet	outer casing	EARTH	23 yellow and green
4	+V	Terminal block (lower)	XT2:0	24V	23 red
5	COM		XT2:2	0V	23 black
6	+5V	Relay module	J1: +5V	5V	23 red
7	COM		J1:AGND	0V	23 black

Table 3.10.8   Wiring table of connection module XS32

S/N	Start Position	End Position		Number of Cable Marker	Line Type
	Connection module XS32	Name	Number		
1	A1	Terminal block (lower)	XT0:0	BAT+	42 red
2	A2		XT0:1	BAT−	42 black
3	A3		XT0:2	A1	12 blue
4	A4		XT0:6	B1	12 blue
5	A5		XT0:3	A2	12 blue
6	A6		XT0:7	B2	12 blue
7	A7		XT0:4	A3	12 blue
8	A8		XT0:8	B3	12 blue
9	A9		XT0:5	A4	12 blue
10	A10		XT0:9	B4	12 blue

Table 3. 10. 9　Wiring table of connection module XS34

| S/N | Start Position | End Position | | Number of Cable Marker | Line Type |
---	Connection module XS34	Name	Number		
1	B1	Terminal row (lower)	XT1:0	KA2	42 red
2	B2		XT1:1	KA1	42 red
3	B3		XT1:2	KB2	42 red
4	B4		XT1:3	KB1	42 red
5	B5		XT1:4	KC2	42 red
6	B6		XT1:5	KC1	42 red
7	B7		XT1:6	KD2	42 red
8	B8		XT1:7	KD1	42 red
9	B9		XT1:8	KE2	42 red
10	B10		XT1:9	KE1	42 red

Table 3. 10. 10　Wiring table of CONV485 interface converter

| S/N | Start Position | End Position | | Number of Cable Marker | Line Type |
---	CONV485 Interface Converter	Name	Number		
1	485 −	Terminal block (upper)	XT0:2	A	12 blue
2	485 +		XT0:6	B	12 blue

- When the battery power display value is ≥90% and <100%, the relay KA, KB, KC and KD are closed, and the LED1, LED2, LED3 and LED4 are on.
- When the battery power display value is =100%, all relays are closed, and all LEDs are on.

Relay description (extended control function)

- The two normally open contacts (KE1, KE2) of the relay have been drawn from the aerial connection to the simulated energy control unit.
- The two normally open contacts of relay KC and KD (KC1, KC2, KD1, KD2) have been drawn through the aerial connection wire to the energy conversion storage control unit.
- The two normally open contacts (KA1, KA2, KB1, KB2) of relay KA and KB have been brought out into the grid-connected inverter control unit through aerial connection wire.

The user can lead these relays to the normally open switches in each system, and connect them to the control circuit of each system in series, so that the working state of each system can be controlled according to the battery power. If the battery power is low, the PV output and the fan output can be connected to the system for the purpose of single charging the battery; when the battery power reaches a certain level, the system can power the DC load or start the grid-connected function; when the battery is full, it will stop charging and so on. Examples are as follows.

The relationship between the battery power (or stimulation battery power) and all output control state

When the battery power detected is <50%, the system performs single charge opera-

Section Ⅲ　Practical Training Project

tion, and only wind power generation system and photovoltaic energy generation system are used to charge the battery.

When the battery power detected is ⩾50% and <70%, the system will charge the battery and power the DC load.

When the battery power detected is ⩾90% and <100%, the system will charge the battery, and power the DC load and the grid-connected inverter.

When the battery power detected is = 100%, the system does not charge the battery, but powers the DC load and the grid-connected inverter.

④ Address settings of DC voltmeter, DC ammeter, multi-function harmonic meter

The address setting method of DC voltmeter and DC ammeter

Press "SET" button to display "PASS" on nixie tube, press "→" key to switch to "ADDr" and then press "SET" button, the nixie tube displays the current address and flashes, change the address by "←" and "→", set the address and press "SET" button to return. Press "→" button to "bAUD" and then press "SET" button, nixie tube shows current baud rate and flashes. Change baud rate by "←" and "→", set the baud rate and press "SET" button to return. Then, switch to "SAVE" by "←" and "→" and press the "SET" button to save and exit.

- PV output voltmeter: communication address (ADDr): 04; baud rate (bAUD): 9600.
- PV output ammeter: communication address (ADDr): 01; baud rate (bAUD): 9600.
- Battery voltmeter: communication address (ADDr): 05; baud rate (bAUD): 9600.
- Battery ammeter: communication address (ADDr): 02; baud rate (bAUD): 9600.
- Inverter input voltmeter: communication address (ADDr): 06; baud rate (bAUD): 9600.
- Inverter input ammeter: communication address (ADDr): 03; baud rate (bAUD): 9600.

The address setting method of multi-function harmonic table

Press "▲" or "▼" in the main menu, select the "user setting", press "confirm" to enter into the password, press "→" to set the password for "0001", click "confirm", select "communication setting" through "▲" or "▼", press "confirm" to enter, find "communication address", and set the address for 07, "baud rate" for 9600 through the "←" and "→".

(2) Configuration design of energy monitoring and management system

Each SymEnergyV2 project is a separate folder. Please try not to edit the folder directly to prevent the project data from being destroyed.

The Project Manager (Figure 3.10.12) provides the function of the creation, deletion, and searching the project directory created.

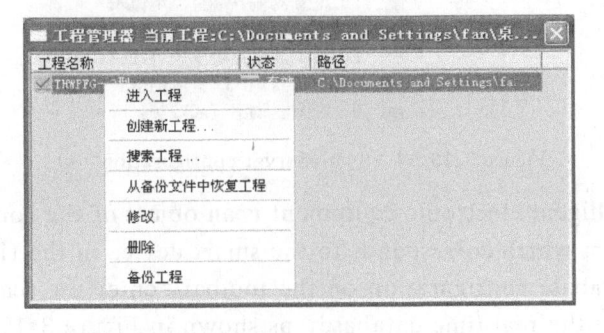

Figure 3.10.12　Project Manager

111

Double-click (or right-click the menu) can set the current project.

① General steps for the preparation of project

• Create a new project in the Project Manager (or search to open a project) and enter the SymEnergyV2 Integrated Development Environment (Figure 3. 10. 13).

Figure 3. 10. 13   SymEnergyV2 Integrated Development Environment

• Click on the I/O communication configuration to enter the SymIoServer configuration (Figure 3. 10. 14 and Figure 3. 10. 15), define the acquisition channel in the acquisition system, create and configure the IED device.

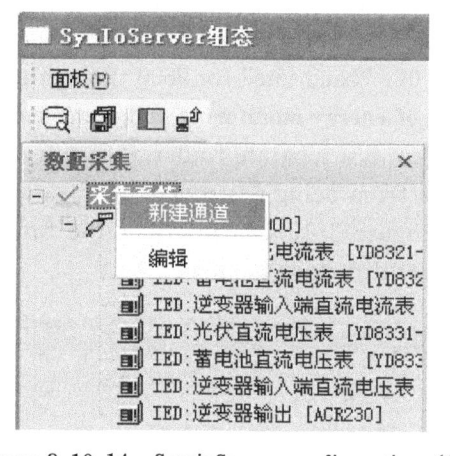

Figure 3. 10. 14   SymioServer configuration (1)

Note：IED intelligent electronic equipment is an object of the communication layer of SymEnergy V2 system which corresponds to the smart device in the field.

• Click the database configuration on the toolbar, enter the real-time database configuration and define the real-time database, as shown in Figure 3. 10. 16.

Real-time database system is the core of SymEnergyV2 platform, and can realize almost all

Section Ⅲ　Practical Training Project

of the advanced functions of SymEnergeV2 in the real-time database layer. A correct understanding of real-time database system is the key to the application of SymEnergyV2 system.

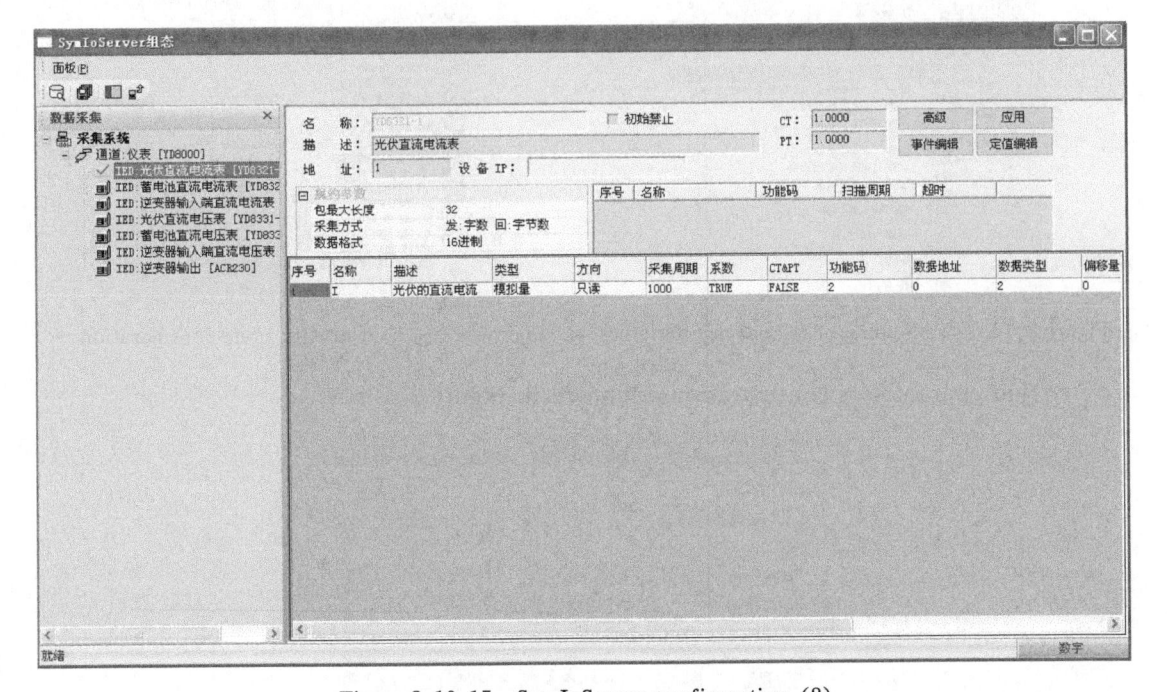

Figure 3.10.15　SymIoServer configuration（2）

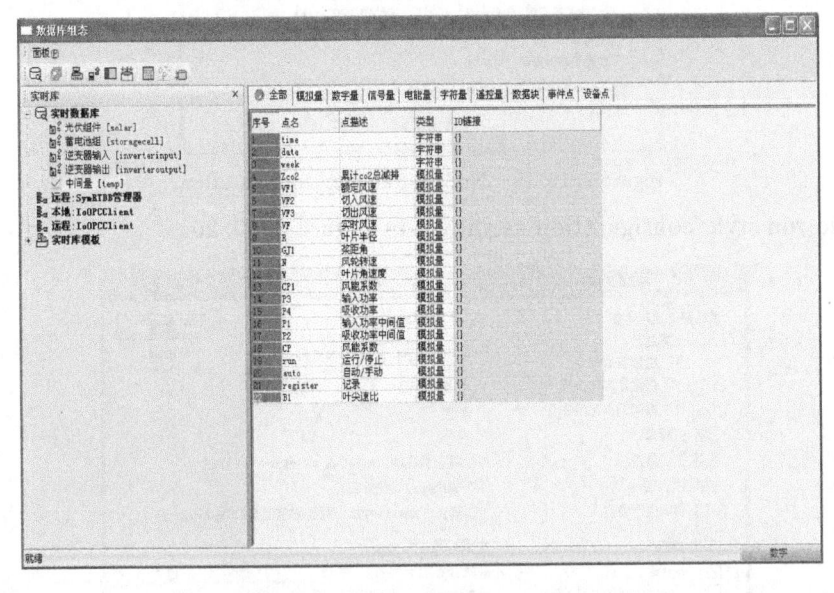

Figure 3.10.16　Real-time database

● Mapping of real-time database configuration and IED device variables，as shown in Figure 3.10.17.

● Network node configuration as shown in Figure 3.10.18.

113

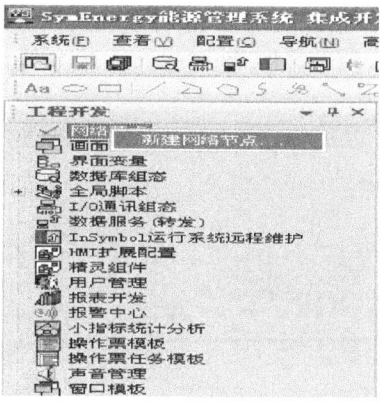

Figure 3.10.17 Mapping of IED device variables Figure 3.10.18 Network node configuration

- Node parameters configuration as shown in Figure 3.10.19.

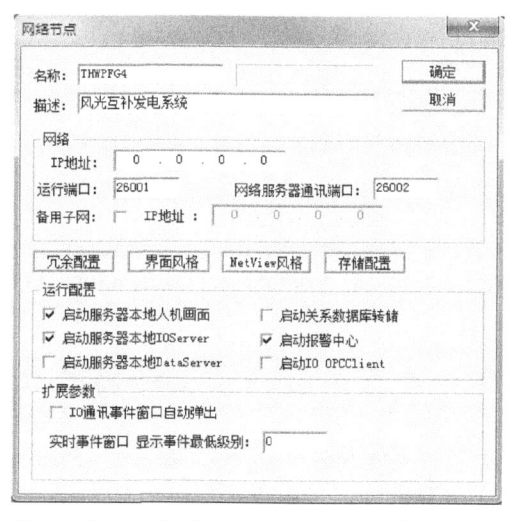

Figure 3.10.19 Node parameter configuration

- Node run style configuration as shown in Figure 3.10.20.

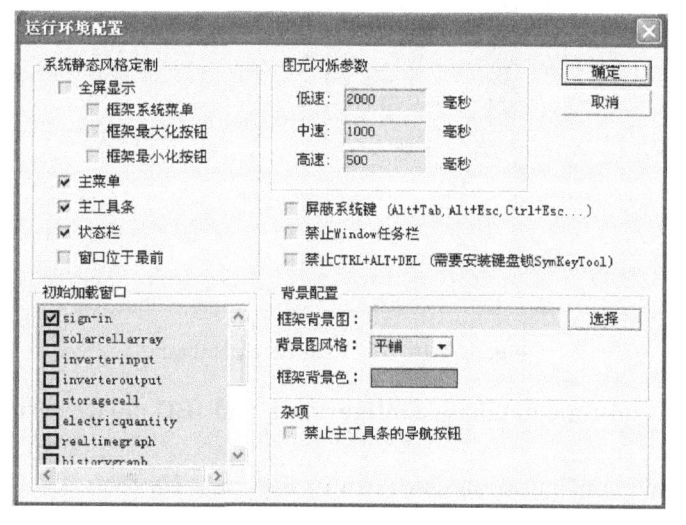

Figure 3.10.20 Node run style configuration

Section Ⅲ   Practical Training Project

● Plan，create and configure windows as shown in Figure 3. 10. 21 and Figure 3. 10. 22.

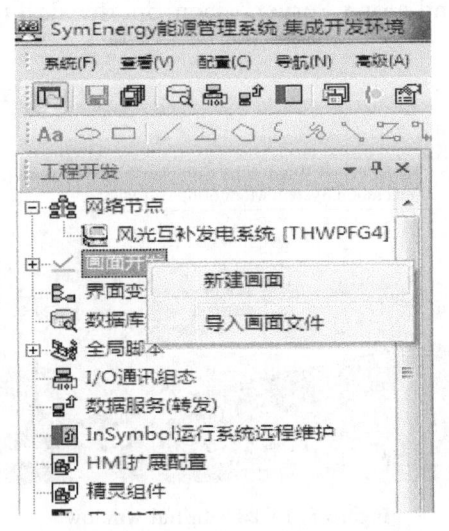

Figure 3. 10. 21   Image development

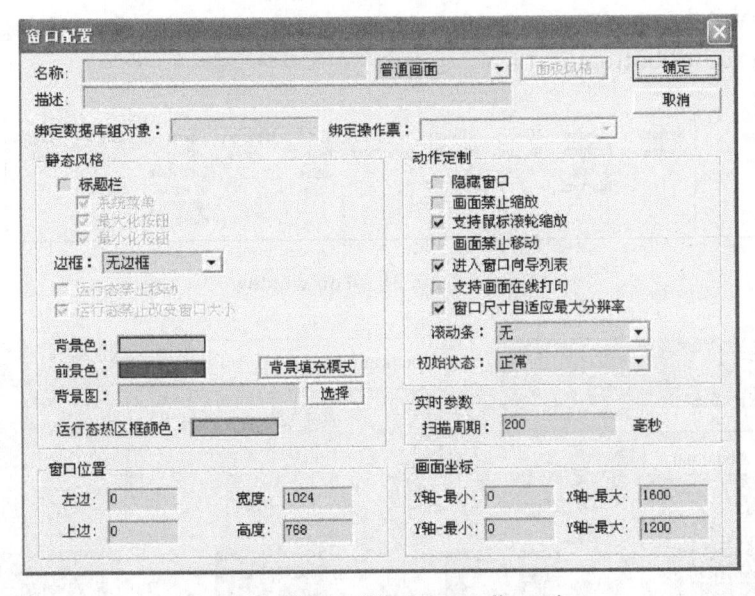

Figure 3. 10. 22   Window configuration

● Image configuration   The SymPower system provides three image configuration modes：

▲ Similar to free graphic drawing and animation configuration of configuration software

▲ Pixel drawing based on valued objects

▲ Pixel drawing based on device objects

② Running system（stand-alone server）   After completing the project in the development system，double-click the "Stand-alone Server" icon on the desktop to run the finished project.

（3）Examples of configuration design

Open the training equipment，connect the aerial plug wire and the communication line，then switch the "main power" of the training platform and the "control power supply" of the control cabinet. Open the touch screen，use the project manager to open the

115

source program of THWPFG-4 Wind-solar Complementary Power Generation System, then double-click the "Stand-alone Server" icon on the desktop and run the project as shown in Figure 3. 10. 23.

Figure 3. 10. 23　Initial window

Click the "Enter" button to enter the monitoring system. Two windows are shown, and they are "Top Window" (Figure 3. 10. 24) and "Communication State Diagram" (Figure 3. 10. 25) from top to bottom in turn.

THWPFG-4 Wind-solar Complementary Power Generation System Monitoring Software	System diagram	Commu-nication state diagram	History Record	History Report	Power parameter	Real time curve	Historical curve	Charact-eristic simulation of wind-driven machine	Log on	Log out	9:54:33 2017/2/7　Tue.

Figure 3. 10. 24　Top window

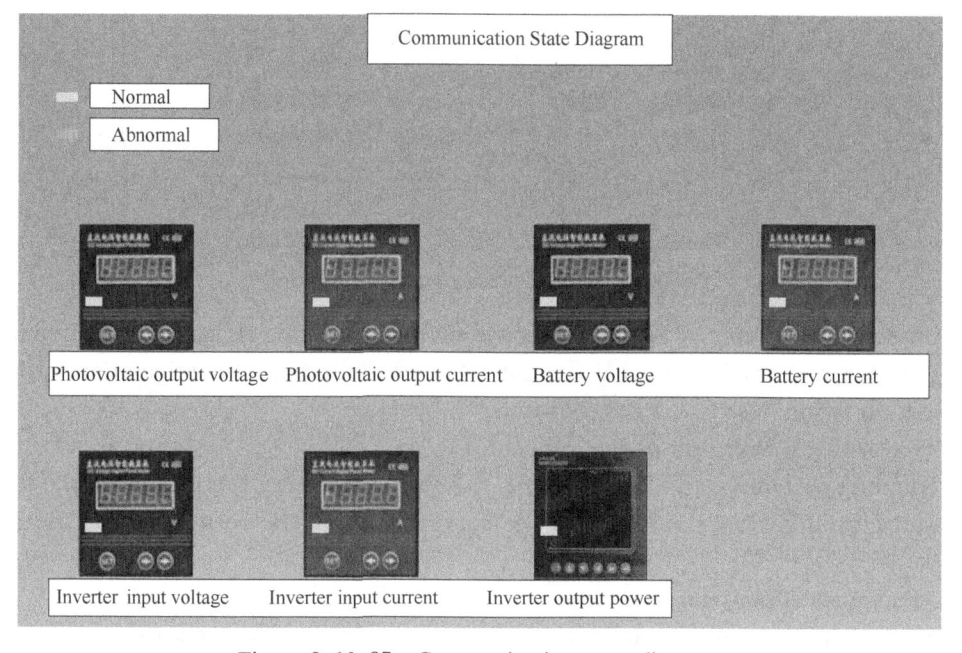

Figure 3. 10. 25　Communication state diagram

Section Ⅲ   Practical Training Project

"Top Window" brings some major operations together including "System Diagram", "Communication State Diagram", "History Report", "Power Parameter", "Real-Time Curve", "History Curve", "Characteristics Simulation of Wind-driven Machine", "Log in" and "Log out".

"Top Window" is always displayed when the system runs. User can click the "communication status diagram" button in any window to quickly return to the "Communication Status Diagram" window. This window can be used to monitor the communication status between monitoring system and DC voltmeter, DC ammeter and AC power meter. If the device communication is normal, the indicator icon of each device is green, otherwise it is red.

Click the "System Diagram", "Communication State Diagram", "History Record", "History Report", "Power Parameter", "Real-time Curve", "History Curve" and "Characteristics Simulation of Wind-driven Machine" button and enter each corresponding window. Click on the "login" to show the login interface (Figure 3. 10. 25). The "log out" button is used to exit the system in the login state, and the user login dialog box is displayed under the un-login state (Figure 3. 10. 26). User name: Admin, password: (no password) and permissions: unlimited.

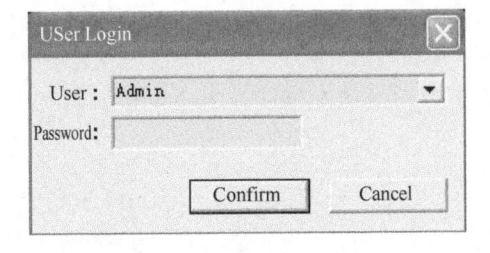

Figure 3. 10. 26   User login

【Project Works】

① Draw the monitoring image in the project monitoring interface.

② Correctly set the parameters of configuration monitoring project and collect data.

117

◆ Reference ◆

[1] Yang Jinhuan. *Application Technology of Solar Photovoltaic Power Generation*. Beijing: Publishing House of Electronics Industry, 2014.

[2] Liu Jing: *Application of Photovoltaic Technology* (*second edition*). Beijing: Chemical Industry Press, 2016.

[3] GB 19565—2017T Pyranometer.

[4] Chen Guocheng. *PWM Inverter Technology and Application*. Beijing: China Electric Power Press, 2007.

[5] Xu Daping. *Principle of Wind Power Generation*. Beijing: Publishing House of Electronics Industry, 2011.